Molecular Biology of Plants

Molecular Biology of Plants
A Text-Manual

Joe H. Cherry

Columbia University Press

NEW YORK AND LONDON 1973

Joe H. Cherry is Professor of Horticulture at Purdue
University, Lafayette, Indiana

Library of Congress Cataloging in Publication Data
Cherry, Joe H 1934–
Molecular biology of plants.
Includes bibliographies.
1. Botanical chemistry—Laboratory manuals.
2. Plant physiology—Laboratory manuals. I. Title.
QK861.C46 581.8′8′028 72-13090
ISBN 0-231-03642-6

Printed in the United States of America

PREFACE

Ten years ago, as a young professor, I set out to teach a new course at Purdue University on plant biochemical and physiological techniques. The course was necessary because many graduate students were not receiving adequate training in experimental technique. In the Agriculture School particularly, students were not gaining sufficient competence and confidence from formal lecture courses in biology and biochemistry to be able to use modern laboratory equipment and follow biochemical procedure. The intent of the course was to provide these students with a wide range of experiments but not necessarily to include all areas of plant physiology. *Molecular Biology of Plants: A Text-Manual* is an outgrowth of those years of teaching. The book is not complete in a sense of covering all areas of plant biochemistry and physiology. The experiments deal mainly with areas of research (e.g., nucleotides, proteins, and nucleic acids) that I have been most interested in during my career. I hope the manual will serve two major purposes—that it will provide an adequate selection of experimental procedures for an advanced laboratory course in plant biochemistry-physiology and also provide the serious student with a reference book relating to those special areas covered in the manual.

Warm thanks go to the many students who have taken my course and have contributed to this manual.

Joe H. Cherry

CONTENTS

Molecular Biology of Plants

PART 1

ENZYMOLOGY

INTRODUCTION

A unique characteristic of a living cell is its ability to permit complex biochemical reactions to proceed rapidly at the temperature of the surrounding environment. The principal agents which participate in these remarkable chemical transformations within the cell are called enzymes. Enzymes are proteins synthesized within the cell which catalyze or speed up thermodynamically possible reactions so that the reaction rate is compatible with the requirements of the cell. The enzyme in no way modifies the free energy (ΔF) of a reaction. Enzymes work at extremely low concentrations. They are generally specific for each reaction; that is, one enzyme does not speed up all reactions, but rather a specific enzyme is required for each different reaction. Being a protein, an enzyme loses its catalytic properties if it is denatured by heat, strong acids or bases, or organic solvents.

PROPERTIES OF AN ENZYME-CATALYZED REACTION

The reaction $A \rightarrow B$ can be thermodynamically described as

$$\Delta F = F_B - F_A$$

If ΔF is negative (−), the reaction is exergonic.

$$\Delta F = F + RT \ln \frac{[B]}{[A]}$$

If ΔF is positive (+), the reaction is endergonic.

Many reactions with positive ΔF proceed at a rapid rate when coupled to an energy source such as ATP. When this occurs the reaction would have a negative overall change in free energy.

Enzyme kinetics should (if possible) be studied under conditions where the rate of reaction is directly proportional to the enzyme concentration. Such a case is shown in Figure 1, which uses the reaction catalyzed by invertase.

The rate of an enzyme-catalyzed reaction increases as the substrate concentration is increased to a maximum level (Figure 2). On saturation with substrate, all catalytic sites of the enzyme are occupied and a maximum velocity (V_m) is achieved. Such reaction kinetics indicate that an enzyme-substrate (ES) complex is formed. A general equation

FIGURE 1. A diagram illustrating an increased reaction rate as a function of enzyme concentration.

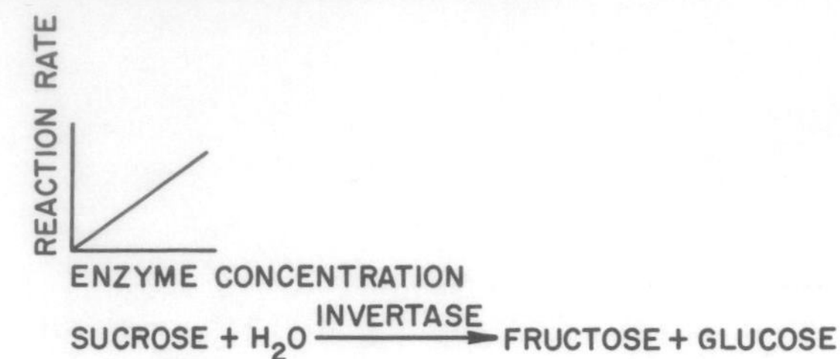

FIGURE 2. The rate of invertase activity as a function of substrate concentration.

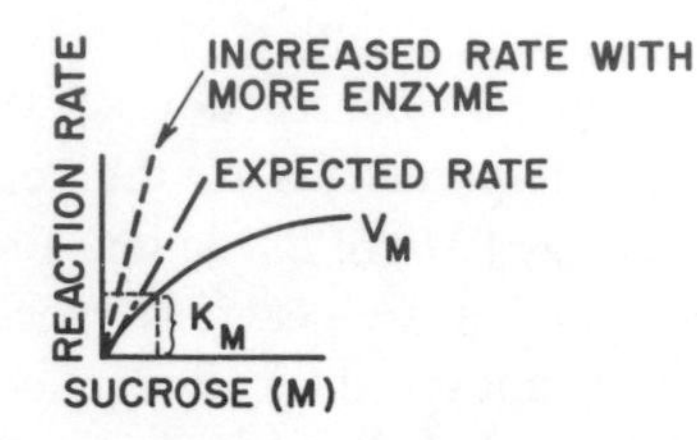

depicting an ES complex and the rate constants for the various reactions are indicated in the equation

$$\text{E (enzyme)} + \text{S} \underset{K_2}{\overset{K_1}{\rightleftharpoons}} \text{ES} \xrightarrow{K_3} \text{E} + \text{P (product)}$$

(The reaction rate depends upon K_3, i.e., K_3 is the rate-limiting step.) The rate of product formation depends on K_3. It is assumed that E + P $\xrightarrow{K_4}$ ES is very small. By definition K_m is equal to the substrate concentration (moles per liter) at half the maximum velocity ($V_m/2$). This constant, K_m, under certain conditions closely approximates the dissociation constant of an ES complex; see the preceding equation.

If we assume that the reaction E + S $\rightleftharpoons$ ES is reversible, then the dissociation constant of ES, defined as K_m, is

$$K_m = \frac{[(\text{E}) - (\text{ES})](\text{S})}{[\text{ES}]}$$

Then, when the equation is rearranged to solve for ES, the result is

$$\text{ES} = \frac{(\text{E})(\text{S})}{K_m + \text{S}}$$

(*Note:* This assumes that K_2 is extremely small.) Now, if the velocity constant of ES is K_3, the measured velocity v is then

$$v = (\text{ES})K_3$$

Substitute ES for

$$\frac{(E)(S)}{K_m + (S)}$$

Thus the equation for v becomes

$$v = K_3\left(\frac{(E)(S)}{K_m + (S)}\right)$$

At maximum velocity (where all enzyme is of the ES form), $V_m = K_3(E)$. Then substitution of V_m for $K_3(E)$ in the equation above yields the Michaelis-Menten equation:

$$v = \frac{V_m(S)}{K_m + (S)}$$

This equation also may be written as

$$\frac{1}{v} = \frac{K_m + S}{V_m(S)}$$

and

$$\frac{1}{v} = \frac{K_m}{V_m} \cdot \frac{1}{S} + \frac{1}{V_m}$$

The Michaelis-Menten equation is illustrated in Figure 3 as the Lineweaver-Burk plot.

HELPFUL HINTS TO THE STUDENT

1. The importance of a "correct" enzyme concentration for a given substrate concentration is shown in Figure 4.
2. In calculating enzyme activity use rate values from the valid (linear) range of the curve.
3. An enzyme unit may be defined as the amount of substrate acted upon per unit time (example: μmoles/min).
4. Specific activity $= \dfrac{\text{units of initial rate}}{\text{mg protein}}$

and this would then be

$$\text{Specific activity} = \frac{\mu\text{moles/min}}{\text{mg protein}}$$

5. (Specific activity) $\times$ (total mg protein) $=$ total enzyme activity

FIGURE 3. Typical Lineweaver-Burk plot.

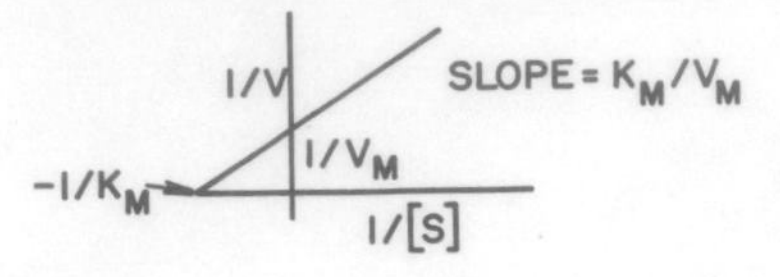

FIGURE 4. Time course of an enzyme-catalyzed reaction, depicting various levels of enzyme concentration.

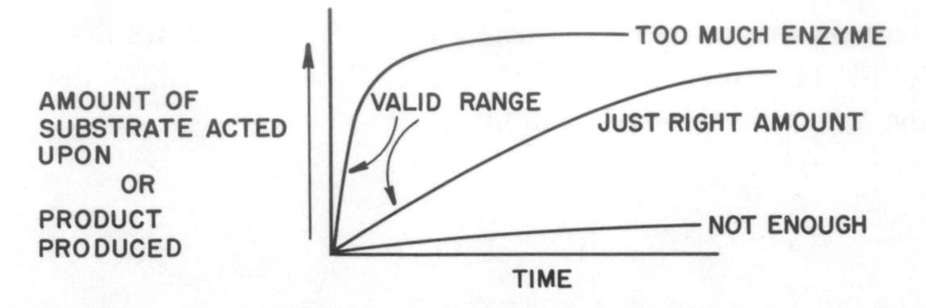

6. Remember to measure initial rates, volume, and protein concentration for each sample.

7. Keep enzyme preparations cold, but during enzyme assay maintain temperature near 30 C or the desired temperature.

CONSIDERATION OF EACH OF THE ENZYME REACTIONS TO BE STUDIED

1. Glucose-6-P Dehydrogenase

The oxidation of G-6-P leads to the production of reduced NADP (NADPH). A characteristic property of the coenzymes, NADP and NAD, is that the reduced forms absorb ultraviolet (U.V.) light near 340 mμ but the oxidized forms do not (Figure 5). (*Note:* To simplify the structure of NADP, many of the H's and OH's not contributing to the basic structure have been omitted. NAD has the same structure as NADP, but the phosphate at the 2′ position of ribose is missing; see Figure 6.)

FIGURE 5. Ultraviolet absorption spectra of oxidized and reduced forms of NAD and NADP.

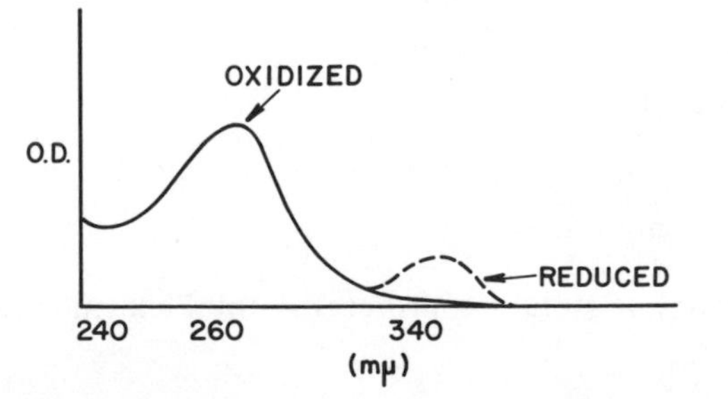

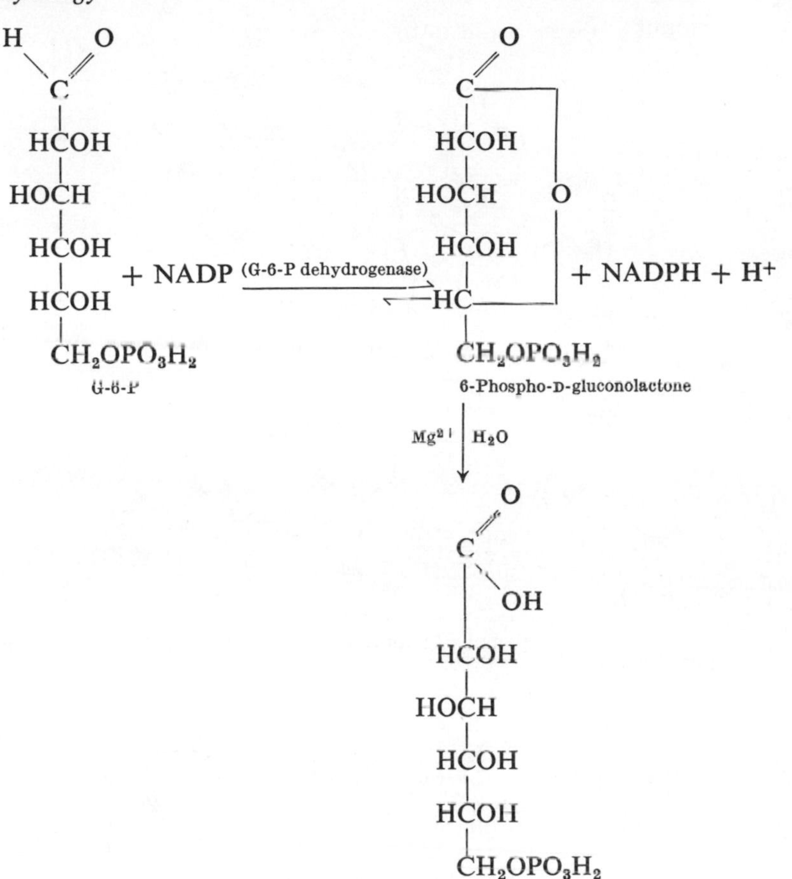

2. Isocitritase

This enzyme participates in the glyoxylate shunt as shown by (*) in the accompanying scheme. The reaction to be studied is the

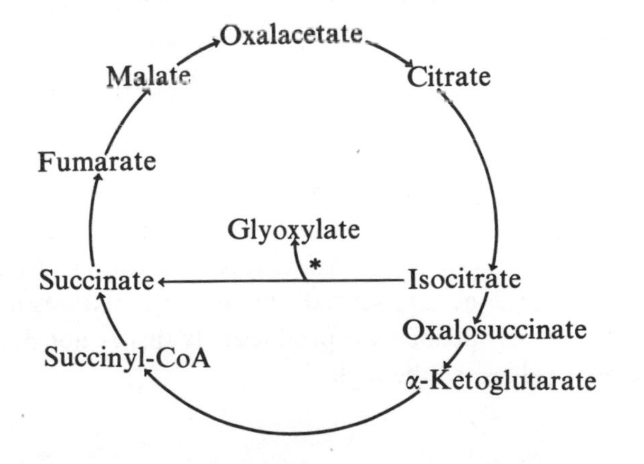

FIGURE 6. The structure of NADP.

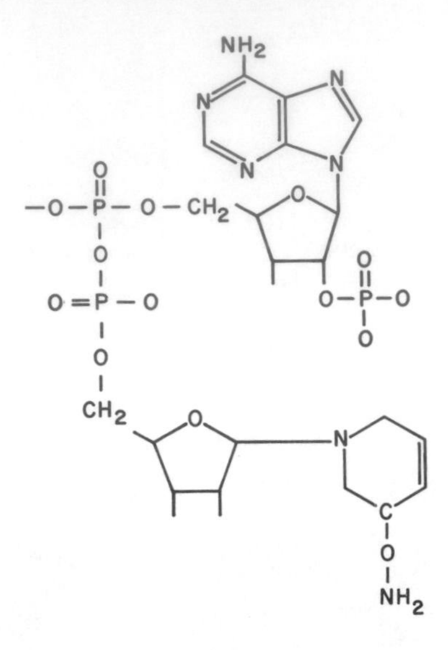

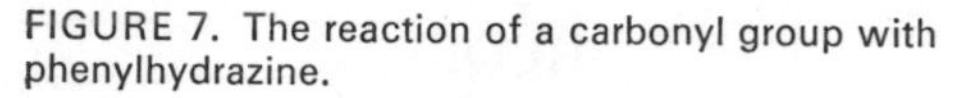

FIGURE 7. The reaction of a carbonyl group with phenylhydrazine.

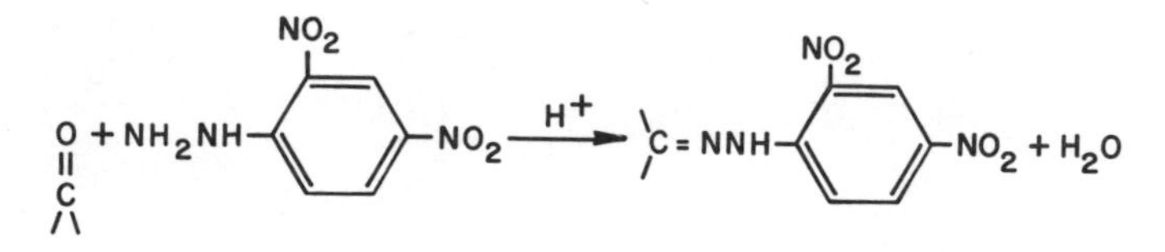

conversion of isocitrate into succinate and glyoxylate as indicated in the equation

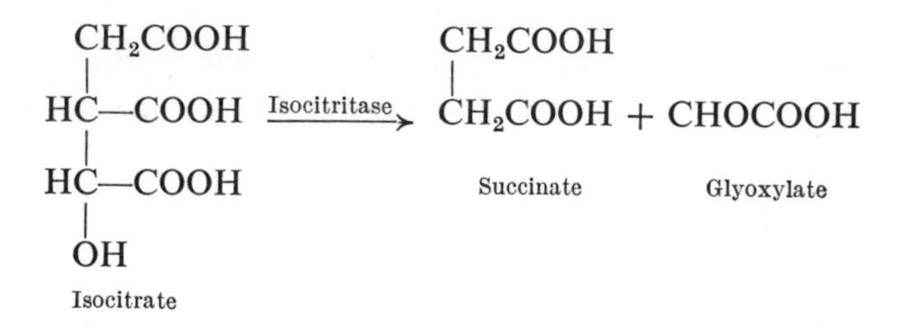

$$\underset{\text{Isocitrate}}{\begin{array}{l} CH_2COOH \\ | \\ HC\text{—}COOH \\ | \\ HC\text{—}COOH \\ | \\ OH \end{array}} \xrightarrow{\text{Isocitritase}} \underset{\text{Succinate}}{\begin{array}{l} CH_2COOH \\ | \\ CH_2COOH \end{array}} + \underset{\text{Glyoxylate}}{CHOCOOH}$$

By the production of glyoxylate, the reaction yields an additional carbonyl group, the formation of which can be measured spectrophotometrically by the phenylhydrazine test (Figure 7). After the glyoxylate phenylhydrazone is produced, the reaction mixtures must be autoclaved to break any Schiff base produced. If this is not done the blank or reference values will be high.

3. *Succinic Dehydrogenase*

This enzyme is localized in mitochondria; therefore a mitochondrial preparation will be used as a source of it.

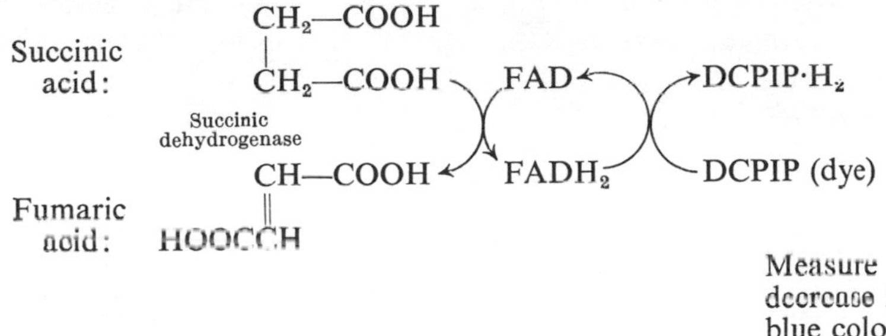

Measure decrease in blue color at 600 mμ

4. *Invertase Activity*

This activity is assayed by measuring sucrose hydrolysis in discs of sugar beet root.

$$\text{Sucrose} \xrightarrow{\text{Invertase}} \text{glucose} + \text{fructose}$$

The enzyme activity is assayed by measuring an increase in reducing sugar (glucose).

5. *Ribonuclease Activity*

Assaying the enzymatic hydrolysis of yeast RNA yields the ribonuclease activity.

$$\underset{\text{(Acid-insoluble)}}{\text{RNA}} \xrightarrow{\text{Ribonuclease}} \underset{\text{(Acid-soluble)}}{2',3'\text{-nucleotides and oligonucleotides}}$$

EXPERIMENT 1 ASSAY OF GLUCOSE-6-PHOSPHATE DEHYDROGENASE ACTIVITY

A. OBJECTIVE

The oxidation of glucose-6-phosphate to 6-phospho-D-gluconolactone requires NADP as a coenzyme. The hydrolysis of the lactone is an exergonic reaction producing 6-phosphogluconic acid. Since (NADPH) absorbs U.V. light at 340 mμ and NADP does not, this system can be used to demonstrate the spectrophotometric measurement of an enzyme-catalyzed reaction.

B. EQUIPMENT AND SUPPLIES

Corn seedlings
Homogenizer
Grinding medium (0.42 *M* mannitol, 0.005 *M* KCl, 0.005 *M* $MgSO_4$, and 0.02 *M* Tris, pH 7.5)
Refrigerated centrifuge
Cheesecloth
Spectrophotometer
Reaction mixture (glucose-6-phosphate, NADP, and KCN)

C. ENZYME REACTION TO BE STUDIED

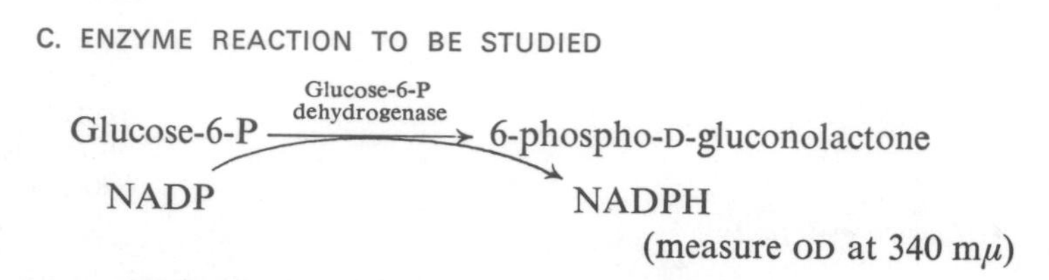

D. EXPERIMENTAL PROCEDURE

Remove 2- to 3-g samples of roots and shoots from 3- to 7-day-old etiolated corn seedlings (germinated in the dark at 29 C) and wash the tissue in cold H_2O. Homogenize each sample of tissue in grinding medium containing 0.42 *M* mannitol, 0.005 *M* KCl, 0.005 *M* $MgSO_4$, and 0.02 *M* Tris, pH 7.5, with a homogenizer for 1 min. Filter the homogenates through four layers of cheesecloth and centrifuge at 500 × *g* for 8 min. Dilute the cleared homogenate to 10 ml with cold homogenizing buffer and use directly as a source of enzyme.

When the enzyme preparation is ready for use, prepare the reaction mixture by adding the indicated volumes of the stock solutions listed in Table 1.

Place the reaction mixture in a water bath at 30 C. When the temperature of the reaction mixture is 30 C, pipet 2.9 ml into each of two quartz cuvettes. To the first cuvette add 0.1 ml buffer; this will be the blank or reference cell. Add 0.1 ml of the plant homogenate to the second cuvette, immediately adjust the spectrophotometer to zero, and

TABLE 1

	Concentration, *M*	*Volume,* ml
NADP	3×10^{-3}	1
Glucose-6-P	9×10^{-2}	1
KCN[a]	4.5×10^{-2}	1
Grinding medium	—	26

[a] *Note:* Do not pipet KCN by mouth.

begin recording any increase in absorbancy at 340 mμ. If the reaction proceeds too slowly or too fast (see Introduction) a larger or smaller volume of plant homogenate can be added and the process repeated. The molar extinction coefficient of NADPH at 340 mμ is 6.22×10^3. Therefore the absorbancy of the reaction mixture when all the NADP is reduced should be near 0.6. From the known molar extinction coefficient the amount of NADP reduced can be calculated.

E. ASSAY OF PROTEIN

Remove 0.5-ml samples with a pipet from each tissue homogenate and precipitate the protein with 1 ml of cold 10% trichloroacetic acid (TCA). Centrifuge at $10,000 \times g$ for 10 min and decant the supernatant. Dissolve the precipitate in 5 ml of 0.1 *N* NaOH and use 0.1-ml aliquots for protein analysis by the method of Lowry *et al.* (1).

1. *Reagents*

(a) 2.0% Na_2CO_3 in 0.1 *N* NaOH
(b) 0.3% $CuSO_4 \cdot 5\ H_2O$ in 1% sodium tartrate
(c) Phenol reagent (Folin-Ciocalteau) 1.0 normal
(d) Standard protein solution: bovine serum albumin (B.S.A.) at 0.2 mg/ml in 0.1 *N* NaOH
(e) Mix reagents (a) and (b) (50:1) shortly before assay

2. *Procedure*

After placing a sample of the NaOH suspension (usually 0.1 ml) in test tubes, add 5 ml of reagent (e) and let stand for 10 min. Then add 0.5 ml of reagent (c) and mix immediately. Allow the sample to stand at room temperature for 30 min and read the absorbancy at 750 mμ. [*Note:* If different volumes are used, adjustments must be made before the addition of reagent (e).]

3. *Standard curve*

Using the solution of B.S.A. (0.1 mg/ml) place portions ranging from 0.1 ml to 0.5 ml (20 to 100 μg) in test tubes and determine the amount of color development.

F. TREATMENT OF DATA

Calculate the total units (1 enzyme unit = change in optical density of 0.01 per minute) and specific activity for all samples assayed. Prepare a short report of your data.

REFERENCE

1. Lowry, O. H., N. J. Rosebrough, A. L. Fair and R. J. Randall. *J. Biol. Chem.* 193: 265 (1951).

General References

Cherry, J. H. *Plant Physiol.* 38: 440 (1963).
Cooperstein, S. J. and A. Lazarow. *J. Biol. Chem.* 189: 665 (1951).
Hiatt, A. J. *Plant Physiol.* 36: 552 (1961).
Ragland, T. E. and D. P. Hackett. *Biochim. Biophys. Acta* 54: 577 (1961).

EXPERIMENT

2 ASSAY OF ISOCITRITASE ACTIVITY

A. OBJECTIVE

Until 1957 a biochemical explanation for the conversion of fat into carbohydrate could not be made. Kornberg and Beevers (1) were the first to show that preparations from castor beans contain the enzymes necessary for the conversion of fat into carbohydrate via acetyl coenzyme A, as in the reaction scheme:

Fat → acyl CoA → acetyl CoA → malate → carbohydrate

Two new enzymes, isocitritase and malate synthetase, are the key enzymes of this new pathway, which is called the glyoxylate cycle. The accompanying diagram describes the cycle. In one turn of the

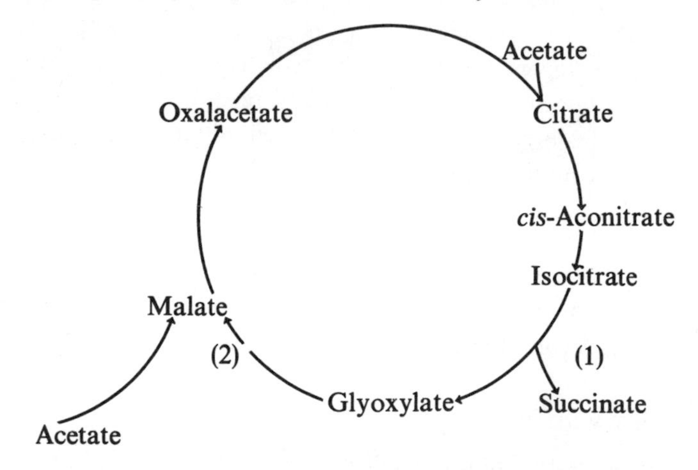

cycle two molecules of acetate are utilized and one molecule of succinate is produced. The succinate can be used in sugar production.

The objective of this experiment is to demonstrate the presence of isocitritase in the fat storage organs of the plant.

B. EQUIPMENT AND SUPPLIES

Pumpkin seedlings (peanut or castor bean seedlings may also be used)
Homogenizer
Grinding medium (0.1 *M* potassium phosphate buffer, pH 7.6)
Refrigerated centrifuge
Cheesecloth
Glyoxylate
Reaction mixture ($MgSO_4$, isocitrate, cysteine-HCl)
Ethanol
Dinitrophenylhydrazine
Trichloroacetic acid (TCA)
Autoclave
Spectrophotometer

C. ENZYME REACTION TO BE STUDIED

$$\text{Isocitrate} \xrightarrow{\text{Isocitritase}} \text{glyoxylate} + \text{succinate}$$

The dinitrophenylhydrazone of glyoxylate can be measured spectrophotometrically at 506 $m\mu$.

D. EXPERIMENTAL PROCEDURE

Isocitritase activities will be determined from homogenates of 4- to 6-day-old pumpkin cotyledon and shoot tissues. Etiolated seedlings grown at 29 C are recommended. Homogenize 5-g samples in 20 ml of cold 0.1 *M* potassium phosphate buffer, pH 7.6, with a homogenizer for 1 min. Filter through cheesecloth and centrifuge at 25,000 × *g* for 15 min. Use the supernatant fraction immediately as a source of enzyme. To assay for isocitritase activity prepare assay tubes, each containing the following: 1 ml of 0.01 *M* potassium phosphate buffer, pH 7.6 (200 μmoles); 0.1 ml of $MgSO_4 \cdot 7\ H_2O$, 13 mg/ml (5 μmoles); 0.1 ml of isocitrate, 42 mg/ml (20 μmoles); 0.1 ml of cysteine-HCl, 7.8 mg/ml (0.5 μmole); 0.4 ml of H_2O. Reactions are started by adding 0.2 ml of the tissue homogenate. After incubation for 15 min at 30 C, stop the reaction by adding 0.1 ml of 80% TCA and placing the tubes in ice. (For each tissue homogenate to be analyzed, two tubes should be incubated at 30 C, and TCA should be added to 2 other tubes which are held in ice.)

Add 1 ml of 0.1% 2,4-dinitrophenylhydrazine in 2 *N* HCl, cover top of each tube with aluminum foil, and then place in an autoclave (15 psi) for 15 min. Cool and then add 2 ml of 95% ethanol. Immediately before reading the samples, add 5 ml of 1.5 *N* NaOH and mix. Read the color intensity at 506 mμ within 3 min after adding alkali. Use 0.4 μmoles of glyoxylate in a total volume of 0.2 ml for a glyoxylate standard (ideally a standard curve should be made). Protein is to be measured by the Lowry (2) method as described in Experiment 1.

E. TREATMENT OF DATA

Calculate isocitritase activity in terms of glyoxylate produced per milligram of protein per hour and write a short report of your findings.

REFERENCES

1. Kornberg, H. L. and H. Beevers. *Biochim. Biophys. Acta* 26: 531 (1957).
2. Lowry, O. H., N. J. Rosebrough, A. L. Fair, and R. J. Randall. *J. Biol. Chem.* 193: 265 (1951).

General References

Beevers, H. *Nature* 191: 433 (1961).
Cherry, J. H. *Plant Physiol.* 38: 440 (1963).
Cooperstein, S. J. and A. Lazarow. *J. Biol. Chem.* 189: 665 (1951).
Gientka-Rychter, A. M. and J. H. Cherry. *Plant Physiol.* 43: 653 (1968).
Hiatt, A. J. *Plant Physiol.* 36: 552 (1961).
Rao, N. A. N. and T. Ramakrishnan. *Biochim. Biophys. Acta* 58: 262 (1962).
Smith, R. A. and I. C. Gunsalus. *J. Biol. Chem.* 229: 305 (1957).

EXPERIMENT 3 ASSAY OF SUCCINIC DEHYDROGENASE

A. OBJECTIVE

Succinic dehydrogenase catalyzes the reversible oxidation of succinate to fumarate when phenazine methosulfate is used as an electron acceptor. In mitochondria and mitochondrial fragments, succinic dehydrogenase is associated with the chain of enzymes and cofactors which transfer electrons from succinate to oxygen by way of cytochrome *c* and cytochrome oxidase, the so-called succinic oxidase system. Flavin and nonheme ferrous iron are covalently bound to the succinic dehydrogenase protein.

The objective of this experiment is to measure the succinic dehydrogenase activity of a mitochondrial preparation. The reduction of flavin adenine dinucleotide (FAD) will be coupled to a dye, dichlorophenolindophenol.

B. EQUIPMENT AND SUPPLIES

Peanut seedlings (5–8 days old)
Mortar and pestle
Grinding medium (0.4 *M* sucrose, 0.1 *M* potassium phosphate, pH 7.1, and 0.001 *M* EDTA)
Refrigerated centrifuge
Cheesecloth
2,6-Dichlorophenolindophenol (DCPIP)
Flavin adenine dinucleotide (FAD)
Phenazine methosulfate (PMS)
Succinic acid
0.06 *M* Tris buffer, pH 7.6
Spectrophotometer

C. ENZYME REACTION TO BE STUDIED

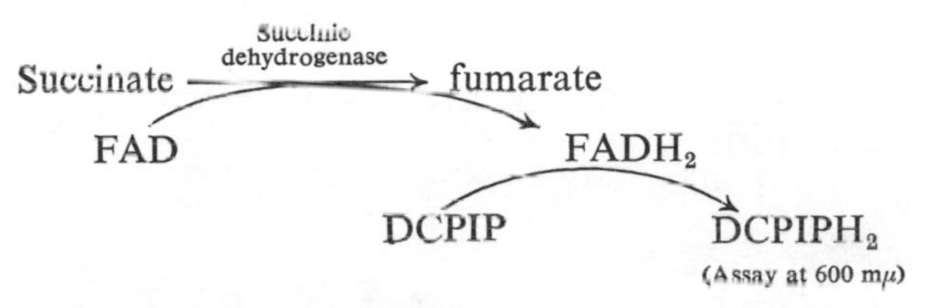

D. EXPERIMENTAL PROCEDURE

Succinic dehydrogenase will be determined by using mitochondria isolated from 6- to 8-day-old peanut cotyledons (etiolated seedlings germinated at 29 C). Grind 10 g of cotyledons in an ice-cold mortar in 10 ml of a solution containing 0.4 *M* sucrose, 0.1 *M* potassium phosphate (pH 7.1), and 0.001 *M* EDTA. Squeeze the homogenate through four layers of cheesecloth and centrifuge at 500 × *g* for 10 min. Decant the supernatant liquid and centrifuge at 20,000 × *g* for 15 min. Suspend the precipitate in 10 ml of 0.06 *M* Tris buffer, pH 7.6 and use directly as a source of enzyme.

To assay succinic dehydrogenase, prepare the reaction mixture by adding the following constituents to glass cuvettes: 0.1 ml of 1.5 *M* potassium phosphate, pH 7.4; 0.1 ml of 1.2 *M* sodium succinate, pH 7.4; 0.1 ml of 9×10^{-4} *M* sodium 2,6-dichlorophenolindophenol (2,6-DCPIP); and 2.5 ml of distilled H_2O. These constituents should previously have been incubated in a water bath at 30 C to ensure proper temperature for the assay. Shortly before assaying the enzyme activity, add 0.1 ml of the mitochondrial preparation, and mix. When all is ready, initiate the reaction by adding 0.1 ml of phenazine methosulfate (PMS) (9 mg/ml) to the cuvette. Measure the rate of reduction of 2,6-DCPIP at 600 mμ during the interval between 15 and 60 sec after initiating. Try to maintain the temperature of the reaction near 30 C.

Measure the protein content of the mitochondrial preparation by the method of Lowry *et al.* (see Experiment 1).

E. TREATMENT OF DATA

Calculate the succinic dehydrogenase activity of your extracts (define the units) and write a short report of your findings.

GENERAL REFERENCES

Cherry, J. H. *Plant Physiol.* 38: 440 (1963).
Hiatt, A. J. *Plant Physiol.* 36: 552 (1961).
Singer, T. P., E. G. Kearney and V. Massey. *Advan. Enzymol.* 18: 65 (1957).

EXPERIMENT 4 ESTIMATION OF INVERTASE ACTIVITY IN SUGAR BEET ROOT DISCS

A. OBJECTIVE

Data from the analysis of invertase activity contributed significantly to the formulation of the first hypothesis accounting for the kinetics of the interaction between an enzyme and its substrate. From data of invertase kinetics, Michaelis and Menten proposed that an enzyme molecule reacts with its substrate to form an enzyme-substrate complex.

Invertase is found in a large number of plant tissues and is often thought to be associated with cell walls. In many storage tissues such as roots and tubers of artichoke, potato, chicory, turnip, carrot, and sugar beet, invertase activity is low in fresh (uninjured) tissue. However, when these tissues are cut into discs and washed in sterile water or buffer, a large increase in invertase activity is noted.

The objectives of this experiment are two. One aim is for the student to learn a technique for measuring enzyme activity without making a homogenate or extract. A second aim is for the student to demonstrate the development of enzyme activity as a function of differentiation (tissue washing).

B. EQUIPMENT AND SUPPLIES

Large sugar beet tuber (other tuber tissues, such as carrot and potato, may be used)
No. 5 cork borer
Cycloheximide
Ethyl acetate
Sucrose-acetate buffer
Spectrophotometer

Nelson's Reagent A. Dissolve 12.5 g of Na_2CO_3 (anhydrous), 12.5 g of potassium tartrate, 10 g of $NaHCO_3$, and 100 g of Na_2SO_4 (anhydrous) in 350 ml of H_2O and dilute to 500 ml

Nelson's Reagent B. Dissolve 7.5 g of $CuSO_4 \cdot 5\,H_2O$ in 50 ml of H_2O and add 1 drop of concentrated H_2SO_4

Arsenomolybdate reagent. Dissolve 25 g of $(NH_4)_6MO_7O_{24} \cdot 4\,H_2O$ in 450 ml H_2O, and add 21 ml of concentrated H_2SO_4. Dissolve 3 g of $Na_2HAsO_4 \cdot 7\,H_2O$ in 25 ml of H_2O and add to the acid molybdate above. Store in brown bottle for 24 hr at 37 C before use

Glucose standard. Dissolve 100 mg of glucose (dextrose) in 1000 ml of H_2O and store in cold room

C. ENZYMATIC REACTION TO BE STUDIED

$$\text{Sucrose} \xrightarrow{\text{Invertase}} \text{Glucose} + \text{Fructose}$$

(Assay increase in reducing sugar by Nelson test)

D. EXPERIMENTAL PROCEDURE

Invertase activity will be determined by incubating discs (1 cm in diameter and 1 mm thick) of the sugar beet root in a medium containing sucrose. Employing a No. 5 cork borer, remove rods of tissue from a sugar beet. Using a razor blade, cut the rod into 1-mm discs. After cutting the tissue into discs, incubate (wash) 50 disc samples in 30 ml of 0.01 *M* phosphate buffer, pH 6.5, at 25 C in a water bath shaker for 0, 6, 12, and 24 hr. A second set of tissue samples will be incubated in 0.01 *M* phosphate buffer containing 10 μg/ml of cycloheximide. Keep discs on ice (wrapped in plastic or waxed paper) until incubated or used for enzyme assay. For clarity, the experimental design of this study is shown in Table 2.

TABLE 2. EXPERIMENTAL PLAN

Sample No.	*Cycloheximide*	*Duration of Washing,*[a] *hr*
1	None	none
2	10 μg/ml	none
3	None	6
4	10 μg/ml	6
5	None	12
6	10 μg/ml	12
7	None	24
8	10 μg/ml	24

[a] Stagger the beginning of the washing process so that all samples will be finished at the same time.

After the tissue has been incubated (washed) for the required times, blot the discs on absorbent paper to remove excess water. Place the discs in a beaker and add about 10 ml of cold ethyl acetate with a small amount (a few grams) of ice. After about 10 min decant the ethyl acetate and repeat the process twice more, followed by a water rinse. The amount of invertase activity present in the ethyl acetate–treated tissue is assayed by incubating 20 discs in 20 ml of a solution containing 0.16 *M* sucrose and 0.045 *M* sodium acetate, pH 5.0 (use a 50-ml flask). Incubate samples for 30 min and 2.5 hr at 30 C. The difference in reducing sugar produced between 0.5 and 2.5 hr may be used to calculate invertase activity. The enzyme reaction is stopped by pipetting 5 ml of the incubation medium (surrounding the tissue) into 5 ml of cold 5% (w/v) Na_2CO_3. Keep these samples cold until assayed for reducing sugar. The amount of reducing sugar is determined by the method of Nelson on an aliquot (1 ml or less) of each of the Na_2CO_3-stopped samples.

Prepare Nelson's alkaline copper reagent by mixing 12.5 ml of Nelson's reagent A with 0.5 ml of Nelson's reagent B (make larger volume if necessary). Add 1 ml (if you use less than 1 ml, make up difference with H_2O) of a solution containing about 100 μg of reducing sugar to a test tube and add 1.0 ml of Nelson's alkaline copper reagent. Shake well and place the tubes in a boiling water bath (a 500-ml beaker or larger if bath is not available) and heat for exactly 20 min. Remove the tubes simultaneously and place them in a beaker of cold water to cool. When the tubes are cool (25 C), add 1.0 ml of arsenomolybdate reagent to each, and shake well several times during a 5-min period to dissolve the Cu_2O and to reduce the arsenomolybdate. After the Cu_2O has dissolved, add 7.0 ml of H_2O to each tube, and mix thoroughly. Read the optical densities in a spectrophotometer at 540 mμ. Use a blank sample (no sugar) to set absorbance to zero. Before determining the reducing sugar of the unknowns, it would be desirable to run a standard curve using glucose. During each assay a glucose standard should also be included.

E. TREATMENT OF DATA

Calculate changes in invertase activity with washing and comment on the effect of cycloheximide. Using your data and other relevant publications, write a paper in the form of a short communication for *Plant Physiology*.

GENERAL REFERENCES

Bacon, J. S. D., I. R. MacDonald and A. H. Knight. *Biochem. J.* 94: 175 (1965).

Cherry, J. H. In *Biochemistry and Physiology of Plant Growth Substances.* Eds. F. Wrightman and G. Setterfield. Runge Press, Ottawa. 1967, p. 417.
Click, E. R. and D. P. Hackett. *Proc. Natl. Acad. Sci. U.S.* 50: 243 (1963).
Edelman, J. and M. A. Hall. *Nature* 201: 296 (1964).
Leaver, C. J. and J. Edelman. *Biochem. J.* 97: 27 (1965).
Nelson, N. *J. Biol. Chem.* 153: 375 (1944).

EXPERIMENT

5 MEASUREMENT OF RIBONUCLEASE (RNase) ACTIVITY

A. INTRODUCTION AND OBJECTIVE

Pancreatic RNase acts on RNA in a manner similar to that of alkali, in that a 2′,3′ diester is transiently formed. In the presence of the enzyme, however, only the C-2′-phosphate bond of the diester is cleaved to yield the 3′ phosphates of the nucleosides. The overall reaction catalyzed by RNase is

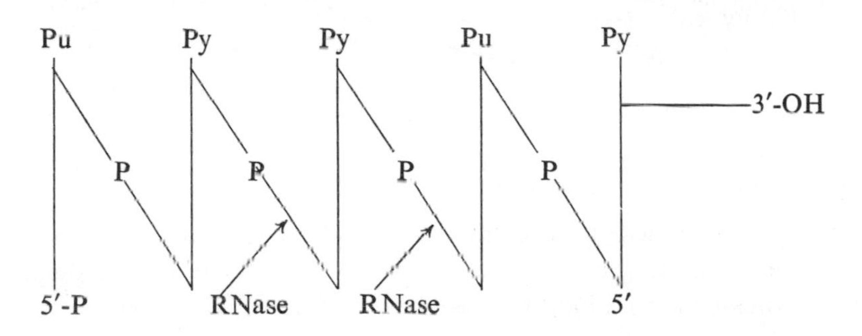

Not all C-5′-P linkages in the RNA strand are attacked by RNase; only the C-3′ diester bonds of pyrimidine (Py) are cleaved.

A number of RNases have been characterized in plants. The specificity of the enzyme for the nucleoside residues in the RNA varies with the plant RNase. Furthermore, in some instances, C-5′-nucleoside phosphates are produced by the action of the RNase. Because of the large amounts of RNase found in plant cells and because the enzyme is very stable, RNase is easy to assay. One objective of this experiment is to correlate relative RNase activities from serial sections of corn roots to the level of maturation.

B. EQUIPMENT AND SUPPLIES

Corn seedlings
Glass homogenizer with Teflon pestle
Phosphate buffer, 0.01 *M*, pH 6.5

Assay medium: 0.5 *M* sucrose, 10^{-3} *M* $MgSO_4$, 10^{-2} *M* KCl, and 2 mg/ml of yeast RNA, pH 6.5.
Perchloric acid–uranyl acetate solutions: 1 *M* $HClO_4$, 2.5×10^{-2} *M* uranyl acetate; 0.2 *M* $HClO_4$, 1.0×10^{-2} *M* uranyl acetate
Refrigerated centrifuge
Spectrophotometer
Glass wool

C. REACTION TO BE STUDIED

Py
P
Pu
P
Pu
P
Py
P
Pu
(acid-insoluble)

$\xrightarrow{\text{RNase}}$ nucleoside monophosphates + oligonucleotides (acid-soluble) + core RNA (nonhydrolyzed)

D. EXPERIMENTAL PROCEDURE

RNase activity in extracts of serial sections of corn roots will be assayed. Excise the primary root of 3-day-old corn seedlings, and divide the apex into three sections (1, 2, and 3) as indicated in Figure 8. Homogenize each sample of root sections (about 2 g) in 10 ml of a 0.01 *M* phosphate buffer, pH 6.5, in an ice-jacketed glass homogenizer using a power-driven Teflon pestle (see Figure 9). Centrifuge the homogenates in a refrigerated centrifuge at $1000 \times g$ for 10 min. Filter the supernatant through glass wool and use the extract directly as a source of enzyme.

RNase activity is assayed by incubating 1 ml of the extract with 1 ml of a solution containing: 0.5 *M* sucrose, 10^{-3} *M* $MgSO_4$, 10^{-2} *M* KCl, and 2 mg/ml of yeast RNA, pH 6.5. Incubate the samples for 30 min at 30 C in a gently shaking water bath. A second identical set of assay tubes should be prepared and kept on ice. At the termination of the incubation period place all tubes in ice and immediately add 0.5 ml of an ice-cold solution containing 1 *M* $HClO_4$ and 2.5×10^{-2} *M* uranyl acetate.

The amount of RNase activity is determined by measuring an

FIGURE 8. Serial sections of a corn root.

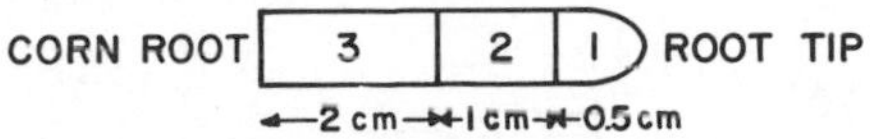

CORN ROOT | 3 | 2 | 1 | ROOT TIP

←2 cm→←1 cm→←0.5 cm

FIGURE 9. An ice-jacketed glass homogenizer with a motor-driven Teflon pestle.

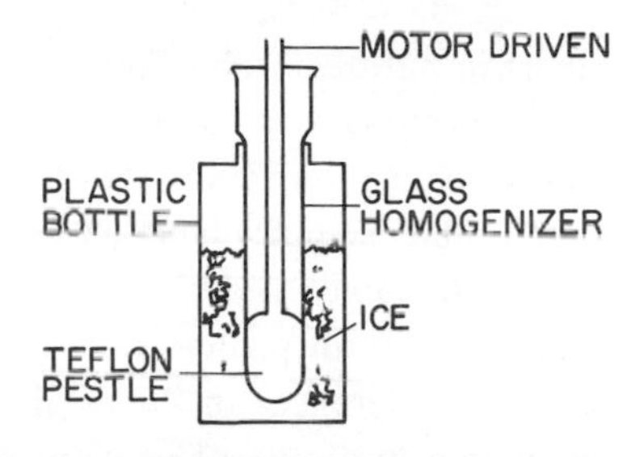

increase in acid-soluble material. Therefore the supernatant is collected by centrifugation at 4000 × *g* for 10 min. Wash the pellet twice by suspending it in 3 ml of 0.2 *M* $HClO_4$ and 1.0×10^{-2} *M* uranyl acetate. After centrifugation add each of the two wash supernatants to the original supernatant and adjust the volume to exactly 10 ml. Using an appropriate blank ($HClO_4$–uranyl acetate), determine the absorbancy at 260 and 290 mμ. Subtract the 290 reading from the reading at 260 and using the relationship that one 260–290 OD unit is equal to 57 μg of RNA, calculate RNase activity.

Determine the protein content of the enzyme extract by the method of Lowry *et al.* as given in Experiment 1.

E. TREATMENT OF DATA

Calculate the total RNase activity for each group of root sections (on a section basis) and also determine the specific activity. Write up your data in the form of a short report.

GENERAL REFERENCES

Barker, C. R. and T. Douglas. *Nature* 188: 943 (1960).
Cherry, J. H. *Biochim. Biophys. Acta* 53: 487 (1962).
Hanson, J. B. *Plant Physiol.* 35: 372 (1960).
Markham, R. and J. J. Strominger. *Biochim. Biophys. Acta* 28: 386 (1958).
Matsuiskita, S. and F. Ibuki. *Biochim. Biophys. Acta* 40: 358 (1960).
Reddi, K. K. *Biochim. Biophys. Acta* 28: 386 (1958).
Wilson, C. M. *Biochim. Biophys. Acta* 76: 324 (1963).
Wilson, C. M. *J. Biol. Chem.* 242: 2260 (1967).

PART 2
ENZYME PURIFICATION

INTRODUCTION

There are many different proteins in nature, each differing in physiological function and chemical properties: for example, the bulk proteins of storage tissue; enzymes which catalyze a variety of chemical reactions; the proteins of small molecular weight which act as specific hormones; and the basic proteins of ribosomes and chromatin. Variations in protein structure allow the diversity in protein function. Such differences in structure are the result of differences in the total number and chemical nature of amino acids in a molecule of protein, and the sequence in which these amino acids are arranged. To understand these distinctions, it is necessary to study the structure of the amino acids and proteins.

PROTEIN STRUCTURE

A protein consists of one or more polypeptide chains which are linear polymers made up of L-amino acid monomeric units linked by peptide bonds. The peptide bond is an amide linkage of the α-amino group of one amino acid to the carboxyl group of an adjacent amino acid. Amino acids differ by the substituent groups on the α-carbon atom. Thus glutamic acid has a $—CH_2—CH_2—COOH$ substituent, lysine a $—CH_2—CH_2—CH_2—NH_2$ group.

Protein structure has been investigated at four levels: the primary, secondary, tertiary, and quaternary structures. Primary structure is the sequence of amino acid residues in a polypeptide chain. Peptide chains vary in length from a few amino acid residues (oligopeptides) to molecules containing 2000 or more amino acids. The first protein to have its sequence determined was the peptide hormone, insulin, in 1955. Since that time the process has been made much less tedious; however, it is still a major undertaking to sequence a protein.

The secondary and tertiary structure of protein is based on forces which impart a specific three-dimensional structure to the polypeptide chain. Many proteins contain helically coiled polypeptide regions; such a region is known as an α-helix. The α-helix, first observed by Pauling and Cory, is stabilized by internal hydrogen bonds formed between amino acid residues in adjacent coils. This helical structure, although probably not characteristic of the entire protein molecule,

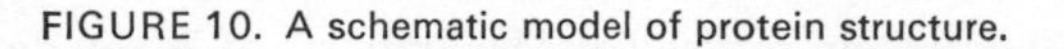

FIGURE 10. A schematic model of protein structure.

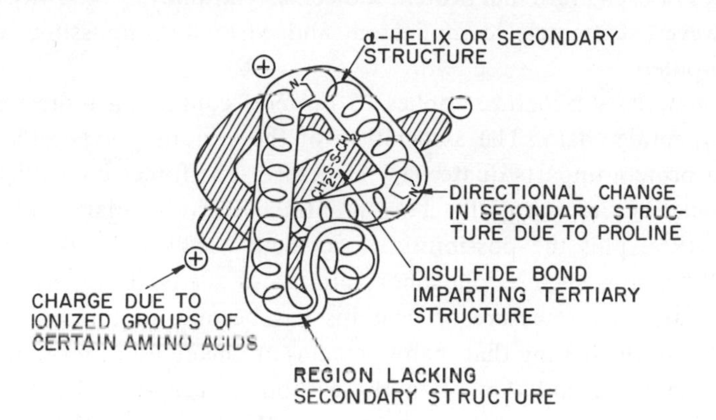

does impart stability to a large part of the molecule. The helix and similar structural forms are called secondary structures of proteins.

The tertiary structure of a protein is the spatial arrangement of the helical coil(s) and the nonhelical regions of the molecule, and is largely a function of the uncombined functional groups of the constituent amino acids. The following factors contribute to the overall tertiary structure of a protein (see the theoretical protein molecule in Figure 10):

1. Disulfide bonds. The sulfhydryl groups of the amino acid cysteine are often involved in intra- or interchain disulfide bonds or both; thus they impart a fixed three-dimensional character to protein molecules.

2. Hydrogen bonds. The phenolic group of tyrosine is frequently involved in hydrogen bonds with the carboxyl groups of glutamic acid and aspartic acid and with the carboxamide group of glutamine and asparagine.

3. Salt bonds. Ionizable functional groups of the constituent amino acids impart structural features to specific proteins.

4. Hydrophobic interactions. Hydrophobic substituents of aliphatic and aromatic amino acids associate because of unfavorable interactions with the water which surrounds and permeates a protein molecule.

5. Proline and hydroxyproline residues. Since proline and hydroxyproline are secondary amines, and not primary amines like the other amino acids, the conformation of proline and hydroxyproline does not fit into an α-helix. Accordingly, a helical structure must be interrupted at each proline or hydroyproline residue in the protein.

Investigation of secondary and tertiary structures of peptide chains is done primarily by X-ray crystallography. This analysis, at its most refined level, allows the assignment of position in three-dimensional space of each atom in a protein molecule. The analysis of X-ray data is, however, extremely complicated and virtually impossible without computers.

Quaternary structure applies to proteins containing more than one polypeptide chain. The association of the various polypeptide chains of a protein imparts quaternary structure. The forces controlling such association are essentially those mentioned for secondary and tertiary structure, plus the possibility of covalent bonds and other covalent modifications such as phosphorylation and adenylation. Investigation of quaternary structure of proteins has become of interest recently, owing to the finding that many proteins are made up of more than one peptide chain and that often these subunits (peptide chains) can be involved in regulation of enzyme activity. The hypothesis that regulation of catalytic activity by enzyme subunits provides a mechanism for feedback inhibition and a possible control of almost every biochemical pathway will provide a basis for additional research.

PROTEIN PURIFICATION

There is no single or simple way to purify all proteins. Procedures useful in the purification of one protein may result in the denaturation of another. Furthermore, slight modifications in a protein may greatly alter its structure and thus affect its behavior during purification and during its general activity. Nevertheless, there are certain fundamental principles of protein purification upon which most fractionation procedures are based. Application of one or more of these procedures can lead to highly purified or crystalline proteins. Several, but not all, of the factors of importance in protein purification are listed below.

1. Solubility properties. The solubility of most proteins in aqueous solutions can be attributed to the hydrophilic interaction between the polar molecules of water and the ionized groups of the protein molecules. Reagents which change the dielectric constants or the ionic strength of an aqueous solution would, therefore, be expected to influence the solubility of proteins. The addition of ethanol or acetone or certain salts such as ammonium sulfate, usually at low temperatures to avoid denaturation, frequently results in precipitation of proteins. Stepwise application of such techniques to heterogeneous protein solutions often results in fractionation (purification) because of the varying degrees of solubility among the proteins in the solution.

Any treatment designed to separate proteins by making use of relative solubility must take into consideration the pH of the medium. Often alteration of only pH under a given set of conditions is sufficient to effect an "isoelectric" precipitation. Choosing the correct condition, pH, ionic strength, or dielectric constant to carry out a given fractionation is somewhat arbitrary and is usually determined by trial and error.

2. Adsorptive properties. Under particular conditions of pH and at low ionic strength, certain proteins will adsorb to various substances. Calcium phosphate gel, DEAE (diethylaminoethyl) cellulose, CMC (carboxymethyl) cellulose, IRC-50 (a carboxyl resin), and other resins are frequently used to adsorb specific proteins from heterogeneous mixtures. The adsorbed proteins can be released from the isolated materials by either altering the pH or increasing the ionic strength. Thus fractions can be obtained by elution of either the extraneous proteins or the desired proteins.

3. Gel filtration. Gel filtration has become an established laboratory technique since the introduction of various Sephadexes. Sephadex and biogels are composed of small particles which contain pores of a known size. When poured into a column, these particles can be used to separate molecules on the basis of size. Molecules larger than the largest pores of the swollen gel, i.e., above the exclusion limit, cannot penetrate the particles, and therefore they pass through the column in the liquid phase outside the particles. Thus they are eluted first. Smaller molecules penetrate the gel particles to varying extents, depending on their size and shape. Molecules are therefore eluted from a gel column in the order of decreasing molecular size. Many Sephadexes and biogels are now available to separate an array of materials from each other. Sephadex ion exchangers (DEAE-Sephadex and CM-Sephadex) and agarose gels are also effective in protein purification.

4. Heavy metal salts of proteins. Addition of heavy metal salts (for example, those of Hg or Pb) to protein solutions often results in precipitation of protein. Such precipitates may be denatured, but under carefully controlled conditions native proteins can be recovered. Such protein–heavy metal salts have proven extremely useful in X-ray crystallographic studies of proteins.

5. Heat denaturation. Since not all proteins are equally unstable when heated in aqueous solution, fractionation can often be achieved by controlled heating. The presence of a substrate will often stabilize a protein against heat denaturation.

6. Protein–nucleic acid complexes. Occasionally during protein purification it becomes necessary to remove basic proteins or contaminating nucleic acid. The controlled addition of sparingly soluble material of opposite charge (for example, addition of basic protein such as

protamine for RNA removal) results in the coprecipitation of the added and unwanted material.

7. Electrophoretic separation (electrophoresis). The net charge on a particular protein is a function of its amino acid composition. Furthermore the charge varies with the pH of the medium in which the protein is dissolved. Accordingly, application of an electric field to be a buffered, heterogeneous protein solution often results in a differential migration of proteins. Thus fractionation can be obtained either in free solution or in heterogeneous systems with an "inert" supporting material, such as starch or polyacrylamine gel, when an electric potential is applied across the material.

ASSAY PROCEDURE DURING FRACTIONATION OR PURIFICATION

Since enzyme (protein) purification involves the selective removal of other proteins. it is necessary to assess the amount of enzymatic activity relative to the amount of protein present. A measure of enzymatic activity per milligram of protein can be employed to indicate the degree of purity of the enzyme in the various fractions obtained during purification. This quantity, the specific activity, can be obtained from a measure of enzymatic activity (true v_0) and a protein determination (for example, the Lowry method).

$$\text{Specific activity} = \frac{\text{units}}{\text{mg protein}}$$

$$= \frac{\mu\text{moles substrate used/min}}{\text{mg protein}}$$

$$= \frac{\mu\text{moles product formed/min}}{\text{mg protein}}$$

Other values of importance are:

$$\text{Total activity} = (\text{specific activity}) \times (\text{total mg protein in preparation})$$

$$\%\ \text{Yield} = \frac{\text{Total activity of given preparation}}{\text{Total activity of the starting material}} \times 100$$

Once values for specific activity, total activity, and yield are known for any given preparation of an enzyme, one can proceed with the refinement of the techniques discussed previously. *It is essential to calculate all three of these values for every fraction obtained during purification.* An increase in specific activity indicates purification. The usefulness of a particular step may then be evaluated with reference to

TABLE 3. RESULTS OF ENZYME FRACTIONATION

Enzyme Fractionation Step	*Volume, ml*	*Protein, mg/ml*	*Total Protein, mg*	*Specific Activity, μmoles product/min / mg protein*	*Total Activity (μmoles product/min) × (total protein)*	*Yield, %*
Original enzyme	1000	1	1000	1	1000	100
First $(NH_4)_2SO_4$ ppt	100	1.143	114.3	7	800	80
Eluant from DEAE cellulose	50	0.428	21.4	28	600	60

the increase in the specific activity of the enzyme, and the yield in the fractions of greatest enrichment. An ideal fractionation would provide complete enrichment (pure enzyme) in 100% yield; in practice, however, such selective procedures are not usually found, and various less ideal fractionation steps are combined. Of course, if enrichment is great, a lower yield may be allowed in a particular step. Table 3 shows a convenient way to present the results obtained during a purification procedure.

EXPERIMENT 6 PURIFICATION OF ISOCITRITASE

A. OBJECTIVE

Isocitritase, an enzyme associated with the glyoxylate shunt, is found in fat-containing seeds (see Experiment 2 for more details). The objective of this experiment is to learn various fractionation techniques normally employed in the purification of many enzymes.

B. EQUIPMENT AND SUPPLIES

Peanut seedlings
$(NH_4)_2SO_4$ (enzyme grade)
Homogenizer
Cheesecloth
Refrigerated centrifuge
DEAE cellulose
Chromatographic column (2.5 × 50 cm)
Glyoxylate
TEM buffer (0.01 *M* Tris, pH 7.6, 0.001 *M* EDTA, 0.005 *M* $MgCl_2$)
Reaction mixture (0.2 *M* KH_2PO_4, 0.005 cysteine-HCl, 0.02 *M* $MgCl_2$, adjust to pH 7.0 with NaOH)

0.2 *M* Isocitric acid	Autoclave
Dinitrophenylhydrazine	Spectrophotometer
25% TCA	

C. EXPERIMENTAL PROCEDURE

1. Enzyme Preparation

Homogenize 100 g of 4-day-old peanut cotyledons with 400 ml of 20% saturated aqueous ammonium sulfate, pH 7.0, in an Omni-Mixer. Squeeze the homogenate through cheesecloth and centrifuge the filtrate for 15 min at 25,000 × *g*. Dialyze the supernatant against 4 liters of distilled water for 4 hr and then remove any precipitate by centrifugation. Adjust the ammonium sulfate concentration to 35% saturation by the slow addition of solid material (see Appendix 1), and after 30 min in the cold collect the precipitate by centrifugation. Dissolve the precipitate in TEM buffer (0.01 *M* Tris-HCl, pH 7.6, 0.001 *M* EDTA, and 0.005 *M* $MgCl_2$), and dialyze for 2–4 hr against the same buffer. Save a small sample for later assay of enzyme activity. Do all purification steps at 4 C.

2. DEAE Cellulose Chromatography

Prepare DEAE cellulose (Whatman grade 23) for column chromatography by the procedure given in Appendix 8. Equilibrate the DEAE cellulose with TEM buffer and pour a column with the ion exchanger to a packed height of 20 cm. Wash the column with 50 ml of TEM buffer and then add the enzyme preparation. Allow the solution to flow through the column without letting the column go "dry." Further wash the protein onto the column by washing with 50 ml of TEM buffer (flow rate should not be more than 1 ml/min). Elute the adsorbed protein with 200 ml of a solution containing TEM buffer and 0.1 *M* NaCl. Collect the eluate in 5-ml fractions and assay each for isocitritase activity. Pool and save about 50 ml of the solution with the highest specific activity and freeze for reassay and calculation of yield and specific activity. This technique should give an enzyme preparation of about fiftyfold purification.

3. Assay

Prepare from stock solutions the reaction mixture which contains in 1 ml: KH_2PO_4, 200 μmoles; cysteine hydrochloride, 5 μmoles; $MgSO_4$, 20 μmoles; and NaOH to adjust to pH 7.0. When ready, add 0.05–0.15 ml of the enzyme fraction or preparation and enough water

TABLE 4

Fraction	*Vol.*	*Protein, mg/ml*	*Total Protein*	*Units of Activity*	*Total Activity*	*Specific Activity*	*Yield %*	*Purification*
Original (20% $AmSO_4$ supernatant)								
25–35% $AmSO_4$ fraction								
DEAE cellulose fraction								

to obtain a final volume of 2 ml. After 10 min preincubation at 30 C, initiate the reaction by adding 0.1 ml of 0.2 *M* isocitrate (20 μmoles). Allow the reaction to proceed for 15 min at 30 C and then stop it by adding 0.5 ml of 25% TCA. A "blank" sample containing water in place of the substrate is put through the identical procedure. Centrifuge the stopped reaction mixture at 10,000 × *g* for 15 min and save the supernatant. Determine glyoxylate in 0.5-ml aliquots of the protein-free supernatant by the method of Smith and Gunsalus as previously described (see Experiment 2, Smith and Gunsalus [1957]). Run a glyoxylate standard curve and calculate the amount of glyoxylate produced by each enzyme fraction. Determine the protein content of each fraction by the Lowry method (see Experiment 1).

D. TREATMENT OF DATA

Present the data in tabulated form in a manner similar to Table 4. Comment on the purification of enzyme in terms of yield and enrichment in the form of a short report.

GENERAL REFERENCES

See references listed in Experiment 2.

Ashworth, J. M. and H. L. Kornberg. *Biochim. Biophys. Acta* 73: 519 (1963).

Carpenter, W. D. and H. Beevers. *Plant Physiol.* 34: 403 (1958).

Cohn, E. J. and J. T. Edsall. *Proteins, Aminoacids and Peptides.* Reinhold, New York. 1948.

Datta, S. P. and A. K. Grzybowski. In *Biochemical Handbook*. Ed. C. Long. Van Nostrand Co., Princeton, N.J., 1961, p. 49.

Dixon, G. H. and H. L. Kornberg. *Biochem. J.* 72: 3 (1957).

Dixon, M. *Biochem. J.* 55: 161 (1953).

Frieden, J. *Am. Chem. Soc.* 80: 6519 (1958).

John, P. C. L. and P. J. Syrett. *Biochem. J.* 105: 465 (1967).

Kornberg, H. L., J. F. Collins and D. Bigley. *Biochim. Biophys. Acta* 39: 9 (1960).
Kornberg, H. L. and H. A. Krebs. *Nature* 179: 988 (1957).
McCurdy, H. C. Jr., and E. C. Cantino. *Plant Physiol.* 35: 465 (1960).
McFadden, B. A. and W. V. Howes. *J. Biol. Chem.* 238: 1737 (1963).
McFadden, B. A., C. R. Rao, A. L. Cohen and T. E. Roche. *Biochemistry* 7: 3574 (1968).
Olson, J. A. *Arch. Biochem. Biophys.* 85: 223 (1959).
Plant, G. W. E. In *Methods in Enzymology*. Vol. V. Eds. S. P. Colowick and N. O. Kaplan. Academic Press. New York. 1962. p. 645.
Syrett, P. J. and P. C. L. John. *Biochim. Biophys. Acta* 151: 295 (1968).

EXPERIMENT 7 PURIFICATION OF LEUCYL-tRNA SYNTHETASE

A. OBJECTIVE

Acylation of tRNA with its respective amino acid is the first step of several reactions in protein synthesis. The reaction requires Mg^{2+}, a reducing agent such as 2-mercaptoethanol, and, of course, the substrates: amino acids, ATP, and tRNA. Because the reaction is so complex and sensitive to phosphatases and nucleases, it is often desirable to purify the synthetase protein. This can be accomplished in a number of ways. However, in this experiment only $(NH_4)_2SO_4$ fractionation and column chromatography techniques are employed.

B. EQUIPMENT AND SUPPLIES

Soybean seedlings (cotyledons)
$(NH_4)_2SO_4$ (enzyme grade)
Grinding medium (0.1 *M* Tris, pH 7.9, 0.04 *M* KCl, 0.04 *M* $MgCl_2$, 0.01 *M* 2-mercaptoethanol)
TMP buffer (0.1 *M* Tris, pH 7.8, 0.05 *M* $MgCl_2$, 2% soluble PVP)
0.01 *M* Sodium phosphate, pH 6.0
Homogenizer
Refrigerated centrifuge
CMC (carboxymethyl cellulose)
Insoluble polyvinyl pyrrolidine (Polyclar-AT)
Glass chromatographic column (2.5 × 50 cm)
^{3}H-L-Leucine, 2 Ci/mmole

tRNA (yeast RNA from General Biochemicals or isolated by methods in Experiment 23)
ATP
2.1-cm Whatman filter paper discs
Insect pins
Hair dryer

C. EXPERIMENTAL PROCEDURE

1. Enzyme Preparation

Germinate soybean seed in a dark, humid environment. After 5 days remove 20 g of cotyledons. With a prechilled mortar and pestle grind the tissue in 20 ml of a solution containing 0.1 *M* Tris, pH 7.9, 0.04 *M* KCl, 0.04 *M* $MgCl_2$, and 0.01 *M* 2-mercaptoethanol. Good tissue homogenization is achieved when the tissue is first ground in a small amount of buffer (0.25 v/w). When the tissue is fairly well ground, add the remaining buffer and complete the homogenization step. At this point, add 20 g of cold Polyclar-AT previously saturated with half-strength homogenization buffer. Stir the Polyclar-AT and tissue homogenate until a homogeneous "dough" is produced. Filter the material through four layers of cheesecloth. After squeezing (with force) the liquid through the cheesecloth, centrifuge out the debris at 20,000 × *g* for 15 min. Save and freeze a sample of the supernatant to be assayed for leucyl-tRNA synthetase activity at a later time. Centrifuge the remaining supernatant fraction at 75,000 × *g* for 1 hr. Dialyze this high-speed supernatant fraction for 3 hr against half-strength homogenizing buffer in the cold. Again centrifuge the dialyzed fraction at 75,000 × *g* for 1 hr. Save a sample (1 ml) of the supernatant for analysis and freeze the remaining portion for further purification.

Thaw the high-speed supernatant fraction and measure the volume. Using the chart given in Appendix 1, slowly (over a 30-min period), add solid $(NH_4)_2SO_4$ to bring the concentration to 40% saturation with respect to $(NH_4)_2SO_4$. It is best to stir the sample slowly, in an ice bath, with a magnetic stirrer. After 30 min remove the protein precipitate by centrifuging the sample at 20,000 × *g* for 15 min. Save the supernatant and again measure the volume. To this fraction add $(NH_4)_2SO_4$ to bring the concentration from 40 to 70% saturation. After 30 min, centrifuge the solution at 20,000 × *g* for 15 min. Discard the supernatant and save the pellet.

Suspend with a glass homogenizer the 40–70% $(NH_4)_2SO_4$ protein pellet in about 4 ml of 0.01 *M* sodium phosphate, pH 6.0. Retain a small portion (0.1–0.3 ml) of the protein suspension for determination of synthetase activity. Freeze or keep the fraction cold until use.

2. CMC Column Fractionation

If not provided, prepare carboxymethyl cellulose (CMC) as described in Appendix 8. Wash and equilibrate the CMC with 0.01 *M* sodium phosphate buffer, pH 6.0. Then pour the CMC slurry into a glass column (2.1 × 50 cm) and pack to a height of 20 cm. Wash the column with about 20 ml of phosphate buffer. Add the 40–70% $(NH_4)_2SO_4$ protein fraction to the CMC column and allow the suspension to flow through the column at a rate of about 2 ml/min. Wash the protein fraction onto the column with small portions of phosphate buffer. Then wash the column with about 200 ml of 0.01 *M* phosphate buffer. Be careful not to disturb the CMC material while adding the protein suspension or phosphate buffer.

Collect the eluate from the column, even the void volume, in 10-ml fractions manually (if a fraction collector is not available). Determine the U.V. absorbancy of each fraction at 280 mμ. Pool the solutions from the four peak tubes and add solid $(NH_4)_2SO_4$ to make to 70% saturation. Collect the protein precipitate and dissolve in about 5 ml of a half-strength homogenizing buffer. Determine the protein content by the method given next, and freeze the sample for measurement of synthetase activity.

3. Estimation of Protein Content

Most proteins exhibit a distinct ultraviolet light absorption maximum at 280 mμ, owing primarily to the presence of tyrosine and tryptophan. Since the tyrosine and tryptophan content of various enzymes varies only within reasonably narrow limits, the absorption peak at 280 mμ has been used by Warburg and Christian as a rapid and fairly sensitive measure of protein concentration. Unfortunately, nucleic acid, which is likely to be present in enzyme preparations, has a strong ultraviolet absorption band at 280 mμ. This acid, however, absorbs much more strongly at 260 mμ than at 280 mμ, whereas with protein the reverse is true. Warburg and Christian have taken advantage of this fact to eliminate, by calculation, the interference of nucleic acids in the estimation of protein.

To use the method of Warburg and Christian, one determines at 260 mμ and 280 mμ the optical density of an appropriately diluted (OD 0.2–2.0) protein solution. Calculate the ratio of 280 mμ/260 mμ and find the factor (*F*) in Table 5. Protein concentration (mg/ml) is

TABLE 5. PROTEIN ESTIMATION BY ULTRAVIOLET ABSORPTION

Ratio at 280 $m\mu/260\ m\mu$	*Nucleic acid, %*	F^a	*Ratio at* 280 $m\mu/260\ m\mu$	*Nucleic acid, %*	F^a
1.75	0.00	1.116	0.846	5.50	0.656
1.63	0.25	1.081	0.822	6.00	0.632
1.52	0.50	1.054	0.804	6.50	0.607
1.40	0.75	1.023	0.784	7.00	0.585
1.36	1.00	0.994	0.767	7.50	0.565
1.25	1.50	0.944	0.753	8.00	0.545
1.16	2.00	0.899	0.730	9.00	0.508
1.09	2.50	0.852	0.705	10.00	0.478
1.03	3.00	0.814	0.671	12.00	0.422
0.979	3.50	0.776	0.644	14.00	0.377
0.939	4.00	0.743	0.615	17.00	0.322
0.874	5.00	0.682	0.595	20.00	0.278

[a] $F \times 1/d = OD_{280}$ = mg protein/ml, where d is cuvette width in centimeters, and OD_{280} is the optical density at 280 mμ. The values are based on those obtained from crystalline yeast enolase.

then calculated from the equation

$$\text{Concentration (mg/ml)} = F \times \frac{1}{w} \times D \times OD_{280}$$

where F is the factor from Table 5, w is the cuvette width in centimeters, and D is the dilution constant.

Another way of obtaining protein content using this general procedure can be calculated from the equation

$$\text{Concentration (mg/ml)} = 1.55\ OD_{280} - 0.76\ OD_{260}$$

where OD_{280} and OD_{260} are optical densities at 280 and 260, respectively.

4. Leucyl-tRNA Synthetase Activity

The filter paper disc method will be employed to assay the leucyl-tRNA synthetase activity of four protein fractions of the purification schedule. Not only will the amount of leucine acylation be determined, but also the general consequence of running the reaction for a long time will be observed. If RNase is more prevalent in some fractions than in others, a loss of charged product will be noted with extended periods of incubation.

The components of the final 1-ml reaction mixture are indicated in Table 6.

TABLE 6

Additives, ml	*Reaction number* 1	2	3	4	5
TMP (0.1 Tris-HCl, pH 7.8, 0.05 *M* $MgCl_2$, 2% soluble PVP)	0.1	0.1	0.1	0.1	0.1
ATP (0.05 *M*, pH 7.8)	0.1	0.1	0.1	0.1	0.1
sRNA (1 mg/ml)	0.5	0.5	0.5	0.5	0.5
^{3}H-L-Leucine, 2 Ci/mmole	0.001	0.001	0.001	0.001	0.001
H_2O	0.2	0.2	0.2	0.2	0.3
Protein fraction	0.1	0.1	0.1	0.1	None

Incubate the tube containing the reaction mixture but without enzyme at 27 C to bring temperature to a constant. The reaction may be started by adding the enzyme to the tubes in sequence at 30-sec intervals. At the end of 3, 6, 9, 12, 18, 25, 30, 45, and 60 min, remove 0.1 ml from the reaction tubes at 30-sec intervals and place each aliquot on individual filter paper discs. The paper discs must be arranged on a Styrofoam board in the order illustrated in the accompanying diagram.

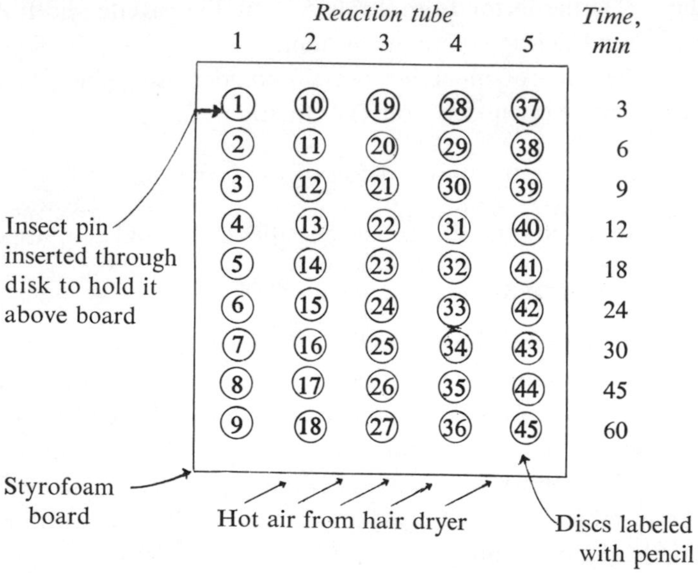

Reaction tube 1	2	3	4	5	*Time, min*
1	10	19	28	37	3
2	11	20	29	38	6
3	12	21	30	39	9
4	13	22	31	40	12
5	14	23	32	41	18
6	15	24	33	42	24
7	16	25	34	43	30
8	17	26	35	44	45
9	18	27	36	45	60

Insect pin inserted through disk to hold it above board

Styrofoam board

Hot air from hair dryer

Discs labeled with pencil

When the reaction has been completed, combine the remaining 0.1 ml left in each tube and remove 0.1 ml on a filter paper disc to determine the amount of radioactivity per 0.1 ml (*do not wash this disc*). Dry all sample discs with a hair dryer. When the discs are no longer moist, incubate them (Nos. 1–45) in the cold in these four solutions:

(a) 10% TCA–0.001 *M* leucine, 30 min
(b) 5% TCA, 30 min
(c) Ethanol-ether (3:1), 30 min
(d) Ether, 15 min

Pour off the ether and dry the disks remaining in the hood under a heat lamp. Then place each disc in a separate scintillation vial containing scintillation fluid, and count the radioactivity in a scintillation counter.

D. TREATMENT OF DATA

From the data, using the initial linear portion of the curve, calculate the pmoles (or counts per minute) of leucine charged per minute per milligram of protein for each of the four fractions. Also present the results in the form of Table 4 (see Experiment 6), giving yield and amount of purification. Write a short note for *Nature* based on your results.

GENERAL REFERENCES

Anderson, M. B. and J. H. Cherry. *Proc. Natl. Acad. Sci. U.S.* 62: 202 (1969).
Keller, E. B. and P. C. Zamecnik. *J. Biol. Chem.* 221: 45 (1956).
Mans, R. J. and G. D. Novelli. *Arch. Biochem. Biophys.* 94: 48 (1961).
Warburg, O. and W. Christian. *Biochem. Z.* 310: 384 (1941).

PART 3
MITOCHONDRIA

INTRODUCTION

Although mitochondria have been known to cytologists for many years, an understanding of their importance as a source of energy for cellular metabolism has come into focus only since about 1930. New isolation techniques and sensitive tools such as the spectrophotometer and the electron microscope have made possible a deeper understanding.

The mitochondria are responsible for cellular respiration and for the oxidation of carbohydrates and fatty acids. In all forms of higher organisms cellular respiration is dependent upon the degradation of pyruvate or acetyl-coenzyme A (a principal substrate) and the concomitant conversion of energy. The respiration process may be defined as the transportation of electrons and protons from substrate to oxygen. The overall general equation for the process is

$$C_6H_{12}O_6 + 6\ O_2 \rightarrow 6\ CO_2 + 6\ H_2O + \text{energy} \qquad (686\ \text{kcal})$$

During the process of electron transport a large portion of the energy derived from the combustion of carbohydrates and fatty acids is conserved as chemical energy in a form suitable for utilization in the cell's functioning. The type of molecule used to conserve the chemical energy varies under different conditions. However, in most endergonic reactions the energy resides in the phosphate bond of adenosine triphosphate (ATP) which is produced in the process of oxidative phosphorylation in the mitochondrion.

PROPERTIES

Mitochondria were first noted by their staining reaction with the dye, Janus Green B. These subcellular organelles are about 7 μ long and appear to be 0.5–1 μ in diameter. Although mitochondria when viewed *in situ* often appear in the "dumbbell" shape, the majority are isodiametric. Since mitochondria are sensitive to their osmotic environment, the highly purified mitochondria may be sphere-shaped.

The mitochondrion is enclosed by a double-layered membrane and therefore is rich in lipids. The internal structure is composed of cristae which appear to be continuous with the intermembrane (Figures 11A and 11B). The cristae are covered with knoblike subunits (electron transport particles) which have been shown to contain the electron

FIGURE 11A. A schematic illustration of a mitochondrion.

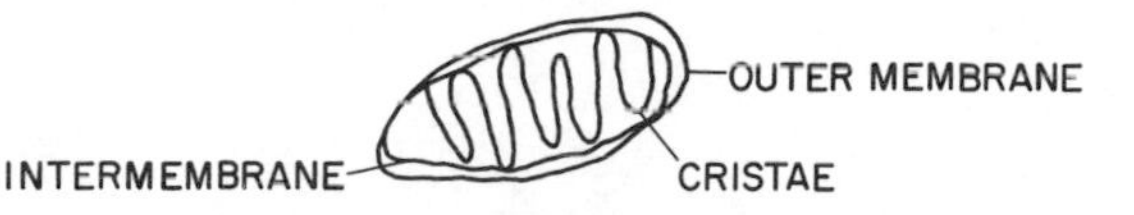

FIGURE 11B. An electron micrograph of a mitochondrial preparation from maize coleoptiles purified on a sucrose gradient. Photograph courtesy of C. A. Lembi and D. J. Morré.

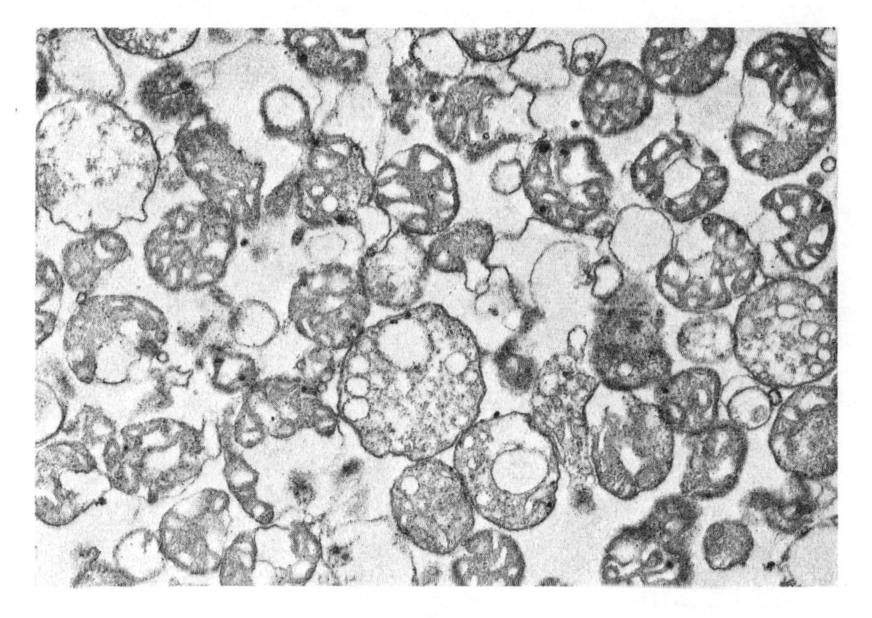

transport system (1). Furthermore, the mitochondrion contains a complement of enzymes which catalyze oxidative and phosphorylative reactions. It also contains the cytochrome system, the necessary electron transport components, and other cofactors to oxidize acetate units to CO_2 and H_2O.

ELECTRON TRANSPORT SYSTEM

During the first 20 to 30 years of the twentieth century many opposing views were generated concerning the nature of cellular oxidation. Keilin (2) was the first to show that the link between the dehydrogenases and the terminal oxidase consists of cytochromes. Cytochromes are complexes of hemes with proteins, collectively termed hemoproteins. The iron (heme) of the cytochromes can be oxidized or reduced. Oxidized cytochromes usually do not absorb visible light, but the reduced forms have specific absorption spectra. Thus spectrophotometric techniques were instrumental in delineating the accompanying scheme of electron transport.

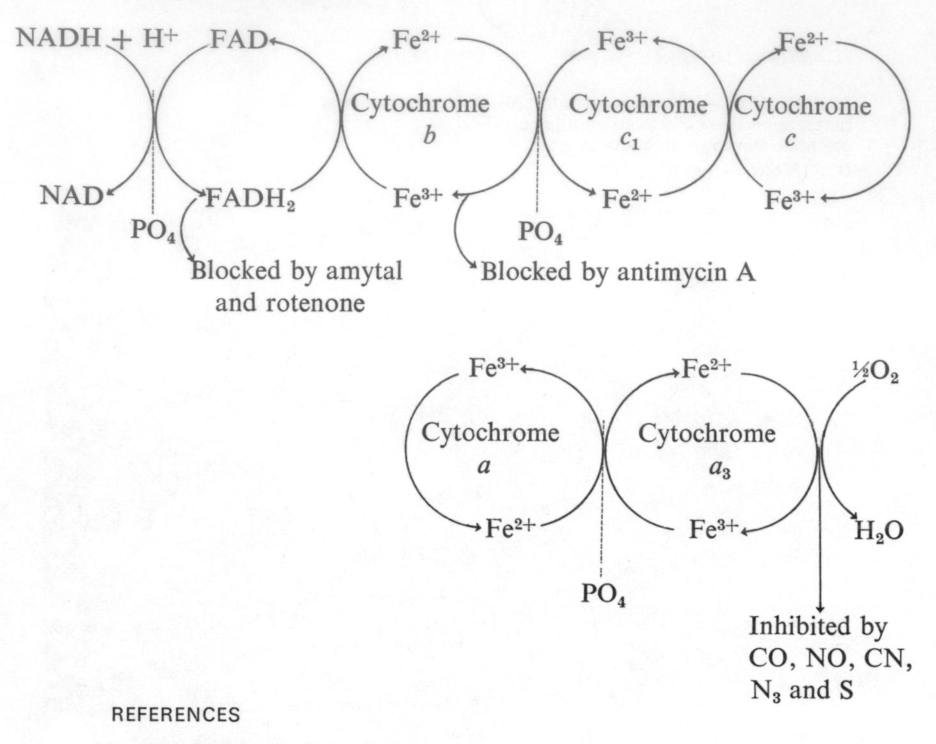

REFERENCES

1. Green, D. E. Fifth International Congress of Biochemistry. Moscow. PL 176 (1961).
2. Keilin, D. *Proc. Roy. Soc. B* 98: 312 (1925).

General References

Bonner, W. D., Jr. In *Plant Biochemistry*, Vol. 7. Eds. J. Bonner and J. Varner. Academic Press, New York. 1965, Chapter 6.

Bensley, D. S. and N. L. Hoerr. *Anat. Rec.* 60: 449 (1934)

Hogeboom, G. H. In *Modern Methods of Plant Analysis*. Eds. S. P. Colowick and N. O. Kaplan. Academic Press, New York. 1955.

Lehninger, A. L. *The Mitochondrion: Molecular Basis of Structure and Function*. W. A. Benjamin, New York. 1964.

EXPERIMENT 8 DETERMINATION OF OXIDATIVE AND PHOSPHORYLATIVE ACTIVITIES OF ISOLATED MITOCHONDRIA

A. OBJECTIVE

In the laboratory an attempt will be made to isolate intact mitochondria and measure their oxidative and phosphorylative activities. Mitochondria are composed of complex protein-lipid systems containing many different enzymes, substrates, and cofactors. As the coupled enzyme reactions of the mitochondria allow for respiratory control, caution should be taken not to disrupt the membrane structure.

When mitochondria are isolated in phosphate, Tris, or many other buffers, several of the cofactors are washed out. Therefore it is necessary to replace these cofactors to obtain respiration.

During the first laboratory session the respiratory activity of mitochondria will be studied using two substrates, succinate and α-ketoglutarate. Pyruvate will also be added to the vessels which contain succinate as a "sparker"—a compound which enhances the oxidation of succinate to fumarate.

The experimental plan requires a Gilson respirometer, because respirometers and various Warburg apparatus are available to most laboratories. However, the student should know that the Clark oxygen electrode is replacing the respirometer and Warburg apparatus. Measurements of oxygen consumption by plant mitochondria with the oxygen electrode are reported by Miller and Koeppe (1).

B. EQUIPMENT AND SUPPLIES

Peanut cotyledons (8-day-old seedlings)
Grinding medium (0.5 *M* sucrose, 0.067 *M* potassium phosphate, pH 6.8, 0.005 *M* EDTA, 1% bovine serum albumin)
Ice-cold mortar and pestle
0.5 *M* Sucrose
Refrigerated centrifuge
Respirometer (or Warburg apparatus)
Reaction mixture (see Table 7)
10% KOH
Reagents for phosphorus determination (Fiske and Subbarow)
Petroleum-lanolin (1:1)
Reagents for determination of nitrogen by Nesslerization
Hydrogen peroxide

C. ENZYMATIC REACTIONS TO BE STUDIED

Phosphorylation coupled to the oxidation of succinate and α-ketoglutarate will be studied with mitochondria isolated from peanut cotyledons.

For succinate the reaction considered is

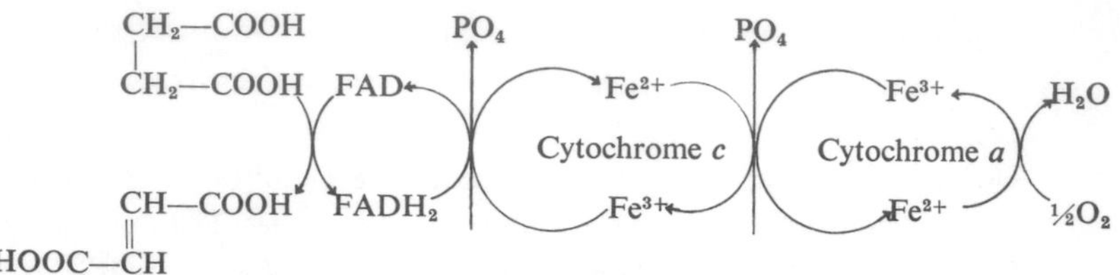

∴ 2 PO_4 esterified
1 atom O used
P/O ratio = 2
(P/O = ratio of the amount of phosphorus incorporated into ATP to the amount of oxygen utilized)

Because malonate and succinate are bound to succinic dehydrogenase in a similar manner, malonate competes for the enzyme and inhibits the oxidation of succinate by a mitochondrial preparation.

Oxidation of α-ketoglutarate is considered in the following reactions:

1. α-Ketoglutarate + NAD + CoA ⟶

$$\begin{array}{l} CH_2\text{—COOH} \\ | \\ CH_2CO\text{—}SCH_2CH_2NHR + CO_2 + NADH + H^+ \end{array}$$

(Succinyl-CoA)

2. Substrate level:

Succinyl CoA + ADP + Pi → succinate + CoA + ATP

3. Cytochrome level:

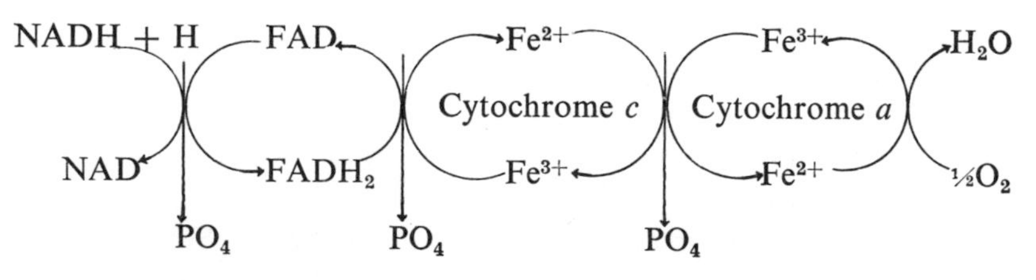

∴ 4 PO_4 esterified
1 atom O consumed
P/O ratio = 4

DNP (dinitrophenol) is known as an uncoupler of oxidative phosphorylation. In the presence of DNP, respiration is increased and the P/O ratio is decreased. Since cyanide binds to cytochrome a_3, it blocks both respiration and phosphorylation. In these experiments both DNP and cyanide will be employed.

D. EXPERIMENTAL PROCEDURE

1. Isolation of Mitochondria

Wash 20 (about 20 g) peanut cotyledons (from 8-day-old seedlings of the Virginia jumbo variety) in cold H_2O. Grind the tissue gently in 3 volumes of a solution containing 0.5 *M* sucrose, 0.067 *M* potassium phosphate, pH 6.8, 0.005 *M* EDTA, and 1% B.S.A. Add about one-third of the homogenizing solution and macerate the tissue in an ice-cold mortar for 5 min. Then add the remaining solution and grind for an additional 2–4 min. Strain the homogenate through four layers of cheesecloth.

All subsequent isolation operations should be performed at 0–5 C. Remove the cellular debris by centrifuging the homogenate at 2000 × *g* for 10 min. Then sediment the mitochondria at 20,000 × *g* for 15 min. Wash the mitochondria once by resuspending the centrifuge pellet in 25 ml of grinding medium and then resedimenting. Suspend the washed mitochondria in 7 ml of 0.5 *M* sucrose in preparation for their addition to the manometer vessels. If two sequential runs are to be made, start with twice the tissue and suspend the washed mitochondria in 14 ml of 0.5 *M* sucrose.

For the student who wishes to pursue this line of research, a procedure has recently been published on the retention of respiratory control of mitochondria from pear fruit (2).

2. Reaction Mixture

The composition of the reaction mixture(s) is given in Table 7.

Each of the two reaction mixtures may be prepared singly in sufficient volume for all the vessels. If other additives, such as dinitrophenol, are added or certain cofactors are omitted, the difference in volume may be adjusted with distilled H_2O.

When the reaction mixture(s) is prepared, keep it cold until used. Prepare the reaction vessels in such manner that an orderly addition of the mixture and other contents can be made. Usually it is helpful to place ice in a large pan and place a second pan over the ice. Then the

TABLE 7. COMPOSITION OF THE REACTION MIXTURE

	Succinate as substrate, ml/vessel	*α-Ketoglutarate as substrate, ml/vessel*
0.1 *M* Potassium phosphate, pH 6.9	0.3	0.3
1.5 *M* sucrose	0.2	0.2
AMP, 10 mg/ml, pH 6.9	0.1	0.1
Cytochrome *c*, 1 mg/ml	0.1	0.1
NAD, 3.3 mg/ml	0.1	0.1
TPP (cocarboxylase), 5.0 mg/ml	0.1	0.1
Coenzyme A, 1 mg/ml	0.1	0.1
0.2 *M* $MgSO_4$	0.1	0.1
1.1 *M* Glucose	0.1	0.1
Hexokinase (solid form)	0.3 mg	0.3 mg
0.2 *M* α-Ketoglutarate, pH 7.0	—	0.2
0.1 *M* Succinic acid, pH 7.0	0.2	—
0.1 *M* Pyruvic acid, pH 7.0	0.1	—
Distilled H_2O	0.5	0.6
DNP (3×10^{-2} *M*)[a]	—	—

[a] If DNP is to be added, substitute 0.1 ml of 2.5×10^{-3} *M* DNP for 0.1 ml of H_2O in the reaction mixture.

vessels may be placed in the pan (Figure 12) and kept cold until placed on the respirometer. While the vessels are being kept cold, prepare them for the experiment in the following manner:

(a) Using a Q-tip or a bit of cotton attached to a toothpick, place a thin layer of a mixture of petroleum-lanolin (1:1, w/w) on the top of the center well (to prevent KOH from spilling out of the well).

(b) Add 0.2 ml of 10% KOH to the center well. (Be very careful not to let the pipet touch anything in the vessel but the inside of the center well.)

(c) Add 2.0 ml of reaction mixture.

(d) Add a small piece of folded filter paper to the center well. Fold the paper so it will open and provide an efficient wick (see Figure 13).

(e) Add 0.5 ml of the mitochondrial preparation. (*Do not touch the wick with the pipet.*)

3. *Respiration*

Respiration will be determined with a Gilson respirometer. Other suitable manometric equipment may be used. The first laboratory period will be devoted to the determination of $QO_2(N)$ (μl of O_2 consumption/mg of mitochondria nitrogen) and P/O ratios of isolated mitochondria employing succinate-pyruvate and α-ketoglutarate as

FIGURE 12. A handy device to keep manometric vessels cold in preparation for an experiment. The bottom pan contains a layer of ice (1 in.); the top pan provides a level surface for the vessels.

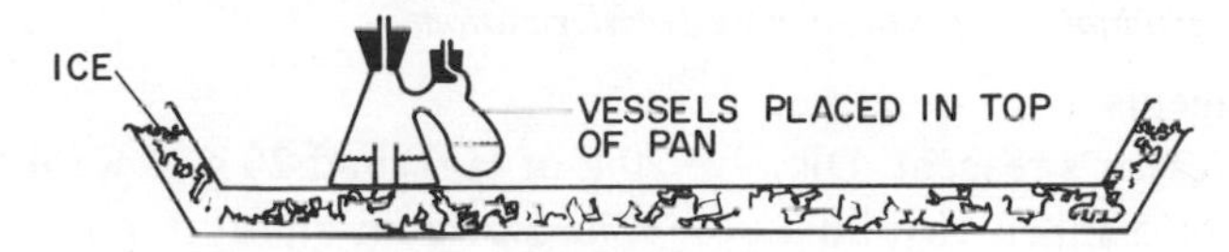

FIGURE 13. A folded filter paper (2.5 × 2 cm) as a wick for the KOH of the center well

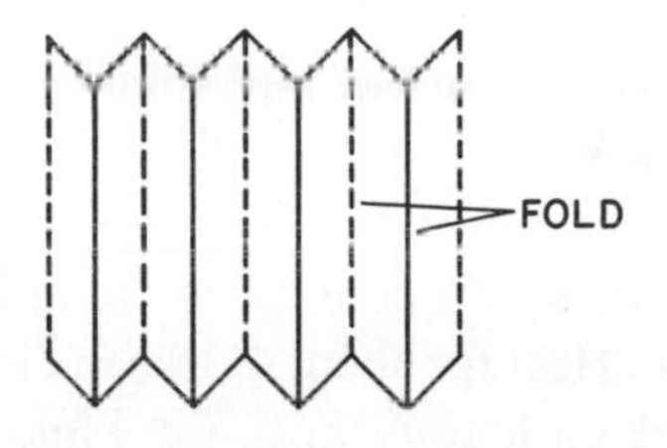

substrates. The mitochondria will be supplied with a complete system of succinate and α-ketoglutarate as substrates but without AMP. The second laboratory period will be devoted to a study of the effect of dinitrophenol and cyanide on respiration and phosphorylation.

Each treatment will require three reaction vessels. When the vessels are ready, add the mitochondrial preparation at 1-min intervals and place each on the respirometer. With the vessels open to the atmosphere (valve open), incubate each one for exactly 5 min. When the first vessel has been incubated for 5 min, close the valve, set the micrometer to zero, and record the reading. Two vessels for each treatment are used to record respiration; the third is removed after 5 min to determine the initial level of phosphorus. Read the consumption of O_2 every 15 min or at shorter intervals. Active respiration usually lasts for about 30 min. Be sure to set up one vessel without mitochondria as a thermobarometer.

4. Inorganic Phosphorus Determination

Phosphorylation coupled to oxidation will be followed by the disappearance of inorganic phosphorus from the reaction mixture. For each treatment one vessel will be used to determine the initial level of Pi, while the other two will be used to determine Pi after respiration. From each vessel remove 1 ml of the reaction mixture and mitochondria and add it to 3 ml of cold 10% TCA. Centrifuge the protein down and analyze the supernatant for Pi by the method of Fiske and Subbarow. Place 0.2 ml of the supernatant (about 0.5 μmole of Pi) into tubes containing 0.4 ml of 10 N H_2SO_4. Add 0.8 ml of 2.5% ammonium molybdate, then 0.4 ml of Fiske-Subbarow reagent.* Dilute with H_2O to 10 ml, mix well, wait 10 min and read at 660 mμ.

5. Determination of Nitrogen by Nesslerization

(a) Reagents

(1) Nessler's reagent. Dissolve 20 g of HgI_2 and 20 g of KI in about 200 ml of distilled H_2O; dissolve 1.75 g of gum ghatti in about 50 ml of H_2O. Mix the two solutions together, dilute to 1 liter, let stand for 24 hr in cold, and filter

(2) 4 *N* KOH

(3) Standard nitrogen source (100 μg N/ml as ammonium sulfate). Use 0.5-ml sample (50 μg) for analysis

(b) Procedure

Add 1 ml of 20% H_2SO_4 to Nessler's tube containing the unknown (0.1 ml of mitochondrial suspension). Heat for about 45 min in digestion rack over hot plate (set temperature on high or hot). Cool and add 3 drops of 30% H_2O_2 and heat on hot plate (highest temperature) for 15 min. Again cool and add 1 drop of H_2O_2 and again heat on hot plate for 15 min. Cool the tubes and add 20 ml of H_2O and 4 ml of 4 *N* KOH. Mix well, and *rapidly* add 2 ml of Nessler's reagent. Dilute to 35 ml with H_2O, mix well, and read at 490 mμ after 10 min.

E. TREATMENT OF DATA

Calculate the QO_2(N) values and P/O ratios of all treatments. Prepare a short report of your data.

REFERENCES

1. Miller, R. J. and D. E. Koeppe. *Plant Physiol.* 47: 832 (1971).
2. Romani, R. and A. Monadjem. *Proc. Natl. Acad. Sci. U.S.* 66: 869 (1970).

General References

Cherry, J. H. *Plant Physiol.* 38: 440 (1963).

Fiske, C. H. and Y. Subbarow. *J. Biol. Chem.* 66: 375 (1925).

Hanson, J. B., A. E. Vatter, M. E. Fisher, and R. F. Bils. *Agron. J.* 5: 295 (1959).

Lanni, F., M. L. Dillion and J. W. Beard. *Proc. Soc. Exp. Biol. Med.* 74: 4 (1950).

* Fiske-Subbarow reagent is prepared by dissolving 0.5 g of 1,2,4-aminonaphthol-sulfonic acid in 5 ml of 20% sodium sulfite with 1–2 ml of 15% sodium bisulfite. When completely dissolved, dilute to 200 ml with 15% sodium bisulfite. Let stand in cold for 24 hr, then filter and store in a brown bottle.

PART 4

PHOTOSYNTHESIS

INTRODUCTION

Photosynthesis is the series of photochemical and chemical reactions which plants use to fix carbon dioxide into carbohydrates while liberating oxygen from water. These reactions occur in higher plant cells in the chloroplasts, which contain not only the photosynthetic machinery, but also DNA, RNA, and a protein-synthesizing system. Biochemists are currently investigating the physiological and evolutionary significance of these findings.

The importance of photosynthesis cannot be overstressed since it is essentially the only means by which carbon is fixed for all biological systems.

PHYSIOLOGICAL STUDIES

Early in the nineteenth century it was established that green plants utilize carbon dioxide to produce sugar and evolve oxygen. Furthermore, it was known that photosynthesis requires light. The quantitative work of de Saussure (1) and others enabled chemists to write the well-known expression for photosynthesis:

$$6\,CO_2 + 6\,H_2O \xrightarrow[\text{Chloroplast}]{\text{Light}} C_6H_{12}O_6 + 6\,O_2$$

Later workers showed that variations in carbon dioxide concentration, light intensity, and temperature all affect the rate of photosynthesis. However, under conditions of excess carbon dioxide and limiting light the rate of photosynthesis is independent of temperature. These results indicated that the photochemical reaction(s) of photosynthesis is temperature-independent. Warburg (2) was then able to show that photosynthesis consists of two broad classes of reactions: light reactions and dark reactions. It was subsequently shown by Emerson and Arnold (3, 4) that the light and dark reactions can be separated in time. Light energy is first absorbed by the chlorophyll of the chloroplast, and this energy is then utilized to fix carbon dioxide by the dark reactions which require a much longer time than is needed to excite chlorophyll.

More recent work on the phenomenon of "enhancement" has led to the discovery of two photoreactions in photosynthesis. Theories on the mechanism by which light energy is collected and funneled to reaction centers have arisen, and many of the photosynthetic electron transport intermediates have been identified and placed in a tentative sequence. The "Z-scheme" attempts to outline the flow of electrons from water to NADPH, the form of reducing power used to fix carbon in the chloroplast (Figure 14).

HILL REACTION

Hill (5) found that isolated chloroplasts, when illuminated, evolve oxygen if a suitable electron acceptor such as ferric iron is present. An electron donor, proved to be H_2O, is oxidized to yield oxygen with the concomitant reduction of Fe^{3+} to Fe^{2+}. As indicated in the Hill reaction

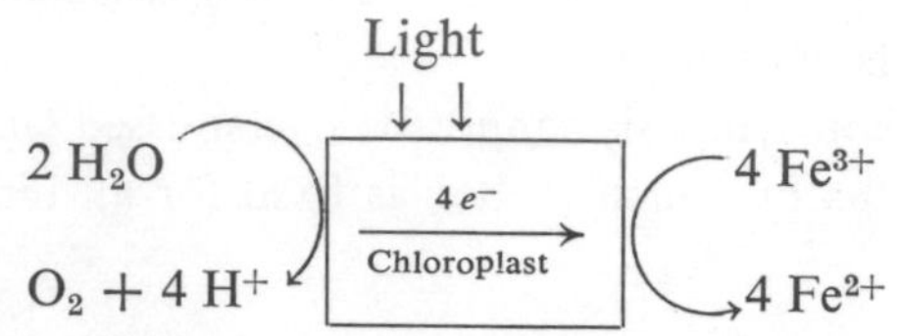

the photolysis of 2 H_2O yields four electrons and four protons which may reduce four Fe^{3+} atoms.

It should be observed that the Hill reaction differs from the photosynthesis reaction in that iron rather than CO_2 is the terminal electron acceptor. Attempts by Hill to prove that CO_2 is the terminal electron acceptor were unsuccessful. However, it has now been shown that light energy absorbed by chlorophyll is used to generate reducing power in the form of NADPH and a high-energy intermediate, ATP. It has also been shown that these compounds are used to fix CO_2 in the dark reaction.

THE BIOCHEMISTRY OF PHOTOSYNTHESIS

The elucidation of the biochemical pathway(s) of CO_2 fixation was not possible until radioactive carbon became available after World War II. Arnon, Allen, and Whatley [see (6)], using an array of ^{14}C compounds, showed that the enzyme machinery for CO_2 fixation exists within the chloroplast. Two pathways of carbohydrate assimilation account for most of the CO_2 fixation in plants. The Calvin (C-3) pathway is found in most plants and is reasonably well understood, even though some steps are not completely elucidated. The "C-4 dicarboxylic acid" pathway is found in maize and most tropical grasses, such as sugar cane. The major labeled products of this pathway are malate and asparate. The differences in the enzymes and reactions of the two pathways are presently being investigated by a number of

FIGURE 14. The electron transport intermediates of the "Z scheme," showing the site of inhibition of DCMU and the proposed sites of ascorbate, dichlorophenol, and $Fe(CN)_6$.

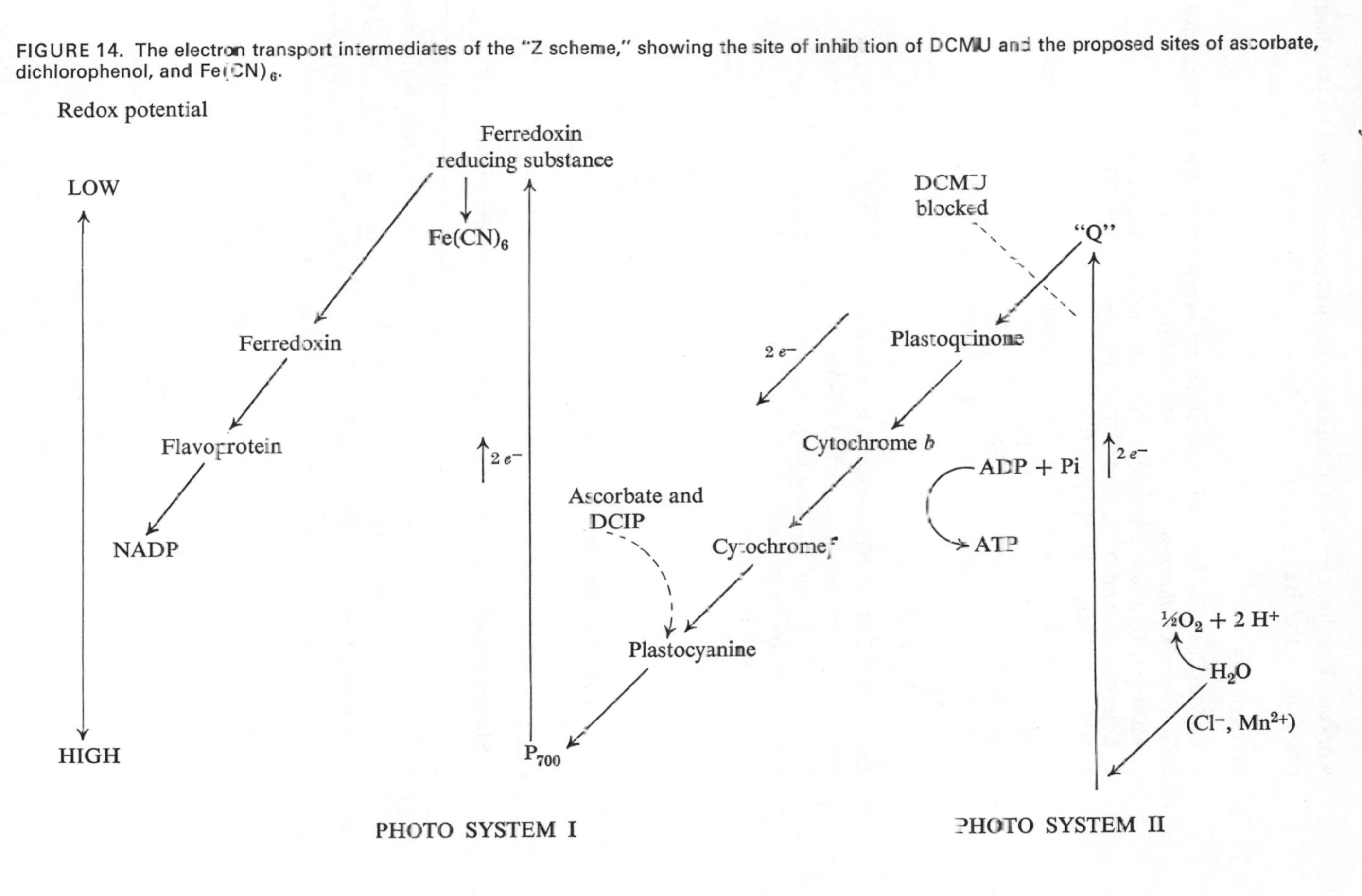

workers. Excellent reviews of this information are presented by Bassham (7), Calvin and Bassham (6), and Hatch and Slack (8).

REFERENCES

1. de Saussure, Th. *Recherches chimiques sur la vegetation*. Nyon, Paris, 1804.
2. Warbury, O. *Biochem. Z.* 166: 386 (1925).
3. Emerson, R. and A. Arnold. *J. Gen. Physiol.* 16: 191 (1932).
4. Emerson, R. and A. Arnold. *J. Gen. Physiol.* 15: 391 (1932).
5. Hill, R. *Nature* 139: 881 (1937).
6. Calvin, M. and J. A. Bassham. In *The Photosynthesis of Carbon Compounds*. H. A. Benjamin, New York. 1962, p. 127.
7. Bassham, J. A. *Ann. Rev. Plant Physiol.* 15: 101 (1964).
8. Hatch, M. D. and C. R. Slack. *Ann. Rev. Plant Physiol.* 21: 141 (1970).

General References

Arnon, D. I., M. B. Allen, and F. R. Whatley. *Nature* 174: 394 (1954).
Blackman, F. F. *Ann. Bot.* 19: 281 (1905).
Blackman, F. F. and A. M. Smith. *Proc. Roy. Soc. B* 83: 389 (1911).
Park, R. B. In *Plant Biochemistry*, Vol. 7. Eds. J. Bonner and J. Varner. Academic Press, New York. 1965.

EXPERIMENT 9 PHOTOSYNTHESIS INHIBITORS

A. OBJECTIVES

Photosynthetic reactions may be studied in whole plants, leaf discs, or isolated chloroplasts. A common way to study the contribution of photosynthesis to the metabolic activity of the plant is through the use of inhibitors. This experiment is designed to measure some of the properties of photosynthesis and to determine the effects of inhibitors on the system. The inhibitors selected for this experiment are herbicides commonly used in weed control.

The herbicide DNBP (4,6-dinitro-*o,sec* butyl phenol) is known to inhibit ATP production by uncoupling oxidative phosphorylation of mitochondria and photosynthetic phosphorylation of the chloroplast. In this experiment the effect of DNBP on photosynthetic phosphorylation using leaf discs will be determined (1).

A second group of herbicides, fitting into the general category of the substituted ureas, strongly inhibits the Hill reaction (see Figure 13). Wessels (2) suggested that one of these substituted ureas, DCMU [3-(3,4-dichlorophenyl)-1,1-dimethylurea], also called Diuron, prevents

FIGURE 15. The structure of Linuron [3-(3,4-dichlorophenyl)-1-methoxyl-1-methylurea].

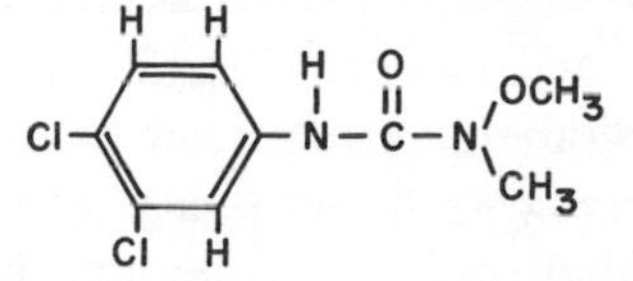

the reduction of cytochrome *f*, an electron carrier in the photosynthetic electron transport chain. Linuron (Figure 15), a substituted urea similar to Diuron, also inhibits the Hill reaction. A second objective of this experiment is to determine the effect of Linuron on the Hill reaction of isolated tomato chloroplasts.

B. EQUIPMENT AND SUPPLIES

- Green tomato plants (30 g leaves)
- Water-bath shaker
- 10^{-4} *M* Citric acid, adjust pH to 6.0 with NH_4OH, 1% sucrose
- Phosphorus, ^{32}P
- 0.05 *M* NaH_2PO_4, pH 6.7
- 0.05 *M* NaH_2PO_4, pH 7.2, 0.35 *M* NaCl
- Gas flow counter
- Homogenizer (Vir-Tis)
- Solutions for Hill reaction
- Refrigerated centrifuge
- Respirometer
- Spectrophotometer
- DNBP, 5 μg/ml
- Linuron, 5 μg/ml in ethanol
- Reaction mixtures (0.037 *M* Tris, pH 7.2, 0.02 *M* KCl, 0.0005 *M* ferricyanide)

C. EXPERIMENTAL PROCEDURE

1. Leaf Discs

Phosphate is freely diffusible across cell membranes, whereas phosphate esterified in ATP is not. This fact will be used to measure photosynthetic phosphorylation of $^{32}PO_4$ into $AT^{32}P$ and the effect of DNBP on this system.

With a No. 5 cork borer punch out about 250 leaf discs from young leaves of tomato plants. Float the discs on ice-cold water until they are used. Place 15 discs, 2.0 ml of 10^{-4} *M* ammonium citrate, pH 6.0, 1% sucrose, and 0.05 μC of ^{32}P-phosphate in each of 16 Warburg flasks, taking care to keep the flasks and solutions cold. Then 0.5 ml of 5 μg/ml DNBP in 10^{-4} *M* ammonium citrate, pH 6.0, and 1% sucrose is added to each of eight flasks while 0.5 ml of only citrate and sucrose is added to the remaining eight flasks. Incubate six flasks containing DNBP and six flasks lacking DNBP in the light (use respirometer lights or other source of about 500 ft-c intensity) at 30 C. Remove two flasks for each sample (from light) at 30, 60, and 120 min. Place the remaining four flasks in the dark and incubate for 120 min. At the termination of the incubation period, pour the incubation solution into the radioactive waste container, wash the tissue with ice-cold 0.05 *M* phosphate buffer, and incubate in ice with the same buffer for 60 min to exchange the adsorbed ^{32}P-phosphate (change phosphate buffer every 10 min). Remove the leaf discs, place in planchets, and dry under heat lamps. Determine the amount of ^{32}P accumulated in one of the duplicate samples of each treatment by using the gas flow counter. Place the other dried samples in scintillation vials containing 15 ml of fluid and count in the liquid scintillation spectrometer. Plot the cpm/leaf disc, using the dark samples as background, and determine the effect of DNBP.

2. Hill Reaction

Homogenize 30 g of washed tomato leaves in 60 ml of 0.05 *M* phosphate buffer, pH 7.2, containing 0.35 *M* of NaCl with a Vir-Tis homogenizer at half speed for 1 min. Filter the solution through six layers of cheesecloth and centrifuge at 1000 × *g* for 10 min. Discard the supernatant solution. Suspend the chloroplast pellet in homogenizing medium to give a concentration of about 0.25 mg chlorophyll/ml (see Section 3, Determination of Chlorophyll).

The Hill reaction will be followed by the reduction of ferricyanide at 420 mμ. To quadruplicate 8-ml portions of a solution containing 300 μmoles of Tris-HCl, pH 7.2, 160 μmoles of KCl, and 4 μmoles of ferricyanide, add 0.1 ml of ethanol containing 0, 0.5, 1.0, 2.0, and 5.0 μg/ml of the herbicide Linuron [3-(3,4-dichlorophenyl)-1-methoxyl-1-methylurea]. Incubate two of the quadruplicate test tubes containing each Linuron concentration for about 10 min at 30 C in darkness. Then add 1 ml of the chloroplast suspension and incubate for 3 min with the lights on, employing a special test tube rack in the respirometer. To the remaining two tubes of each Linuron concentration add 1 ml of chloroplast suspension but keep in the ice bath in the dark. At the end of the incubation period, turn the lights off and place the tubes in ice, add 1 ml of 20% TCA, and mix the contents well. Centrifuge the tubes at 3000 × *g* for 10 min to remove the protein, and read the supernatant at 420 mμ to determine the reduction of ferricyanide. Calculate the change in OD_{420}/min/mg chlorophyll after subtracting the average of the tubes incubated in the dark from the average of those in the light for each treatment.

3. Determination of Chlorophyll

The concentrations of chlorophyll a and b are determined as described by Arnon (3), who measured the optical density of 80% aqueous acetone extracts with a Beckmann spectrophotometer at 645 and 663 mμ in a 10-mm cell. Simultaneous equations using the specific absorption coefficients for chlorophyll a and b given by MacKinney (4) will be used.

The following procedure will be used for chlorophyll determinations: Shake 0.5 ml of the chloroplast suspension with 4.5 ml of water and 20 ml of acetone in a glass-stoppered volumetric flask. Centrifuge or filter the suspension and determine the absorbancy at 645 and 663 mμ. Total chlorophyll content may be determined by substituting the absorbancy (OD) values into the equation which takes into account the dilution factor and gives chlorophyll concentration, C, in milligrams per milliliter:

$$C = (20.2A_{645} + 8.02A_{663}) \times \frac{25}{1000 \times 0.5}$$

D. TREATMENT OF DATA

Prepare graphs showing effect of DNBP on uptake of ^{32}P with time and the effect of Linuron on the Hill reaction. Write a short report for *Science*.

REFERENCES

1. Wojtaszek, T., J. H. Cherry, and G. F. Warren. *Plant Physiol.* 41: 34 (1966).
2. Wessels, J. S. C. *Proc. Roy. Soc. B* 157: 345 (1963).
3. Arnon, D. I. *Plant Physiol.* 24: 1 (1949).
4. MacKinney, G. *J. Biol. Chem.* 140: 315 (1941).

General References

Colby, S. R. Ph.D. thesis, Purdue University, 1963.
Good, N. E. *Plant Physiol.* 36: 788 (1961).
Herbicide Handbook of the Weed Society of America. W. F. Humphrey Press, Geneva, N.Y. 1967 Edition.
Moreland, D. E. and K. L. Hill. *Weeds* 10: 229 (1962).

PART

5

NUCLEOTIDES

INTRODUCTION

Nucleotides constitute one of the most active, varied, and versatile groups of natural compounds. Polymers of nucleotides, when combined in a specified sequence, make up DNA, the genetic material of the cell, as well as RNA, the translated form of the genetic information. Nucleotides participate in and regulate all phases of metabolism; for example: carbohydrate (ATP, UTP); lipid (CTP); protein (GTP, ATP); nucleic acid (all nucleoside triphosphates).

Many of these processes take place only when the substrate is "activated" by combination with a nucleotide. In addition, nucleotides have a major role in the efficient use of energy in living systems. Nucleotide derivatives such as NAD, NADP, and coenzyme A are instrumental in oxidation-reduction and 2-carbon transfers.

The first nucleotide discovered was inosinic acid, a deamination product of AMP, which was isolated from meat by Liebig in 1847. ATP was discovered in 1929 by Lohman in Germany and by Fiske and Subbarow in the United States. Today there are at least 110 known naturally occurring nucleotides.

NUCLEOTIDE COMPOSITION

Nucleotides are composed of a nitrogenous purine or pyrimidine base, ribose or deoxyribose, and a phosphate esterified to the third or fifth carbon of pentose. The components shown in Figures 16 and 17 make up the many nucleotides.

The nucleoside di- and triphosphates are formed by the addition of one or two phosphate groups through pyrophosphate linkage to the monophosphate (see the structure of ATP in Figure 18).

PURINE SYNTHESIS

The initial isotopic approach to the study of purine synthesis was carried out by Buchanan and co-workers (1, 2), who administered radioactive compounds to pigeons. Uric acid was isolated from the excreta. The radioactive materials administered to the pigeons included relatively simple compounds such as lactate, pyruvate, acetate, glycine, formate, and CO_2. Yet glycine, formate, and CO_2 proved to be the critical building blocks in the early stages of the biosynthesis of the

FIGURE 16. Structures of the bases common to most nucleotides.

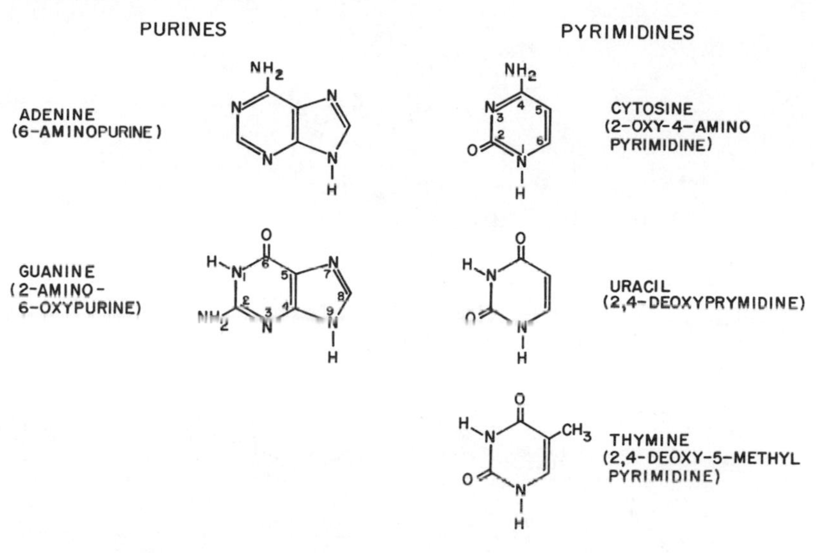

FIGURE 17. Structures of the two sugars of nucleotides. The structure of 5′-AMP is given as an example of nucleotides.

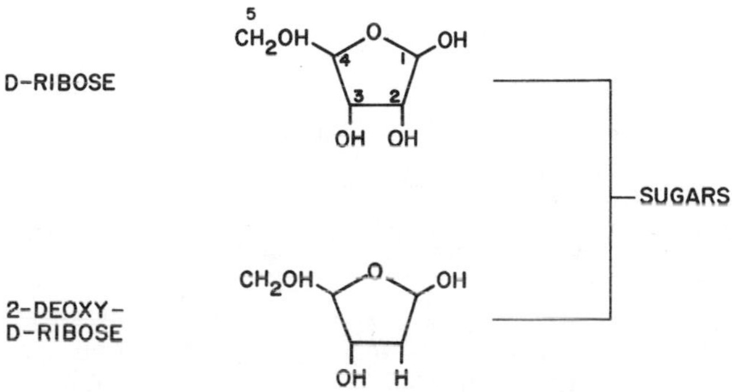

5′-AMP

NUCLEOTIDE

NITROGEN BASE + SUGAR = NUCLEOSIDE

FIGURE 18. The structure of adenosine triphosphate (ATP).

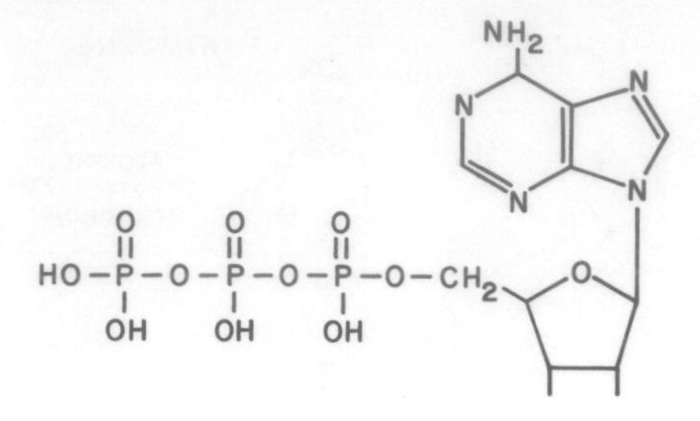

FIGURE 19. The structure of uric acid, showing the origin of C and N members of the ring.

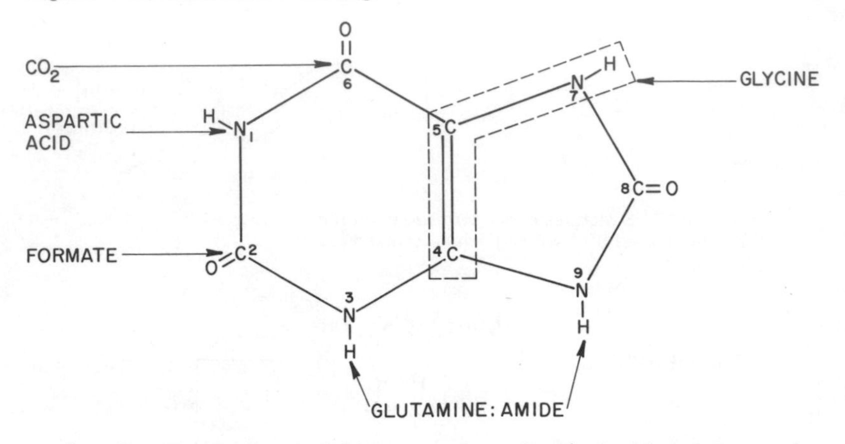

purine ring. Buchanan and his co-workers (1, 2) elucidated the purine biosynthesis pathway by administering one labeled compound at a time and determining the exact localization of the isotope in specific atoms of uric acid. On the basis of these preliminary studies with labeled precursors of uric acid it was shown that carbon atoms 2 and 8 came from formate, carbon atom 6 came from CO_2, and carbon atom 4 came from carbon atom 1 of glycine. Other studies have shown that nitrogen atom 7 of uric acid also came from glycine. The results of these early studies are portrayed in Figure 19.

Later Greenberg, using pigeon liver homogenates, showed that radioactive formate and CO_2 are incorporated into hypoxanthine. Greenberg and other workers in the field (3–6) further showed that hypoxanthine is not a primary product but is derived from inosinic acid, the ribonucleotide of hypoxanthine. Inosinic acid was well known as the deamination product of adenylic acid, and in this early stage of nucleotide work there was no reason to believe that it was a precursor of adenylic acid. However, attention was focused on the role of inosinic acid as the crossroads of purine metabolism. After these early experiments, Buchanan and Greenberg and their respective co-workers isolated the individual enzyme systems and then determined the intermediate products involved in the *de novo* synthesis of the purines. A good description of the various reactions is summarized by van Potter (7).

PYRIMIDINE SYNTHESIS

Orotic acid is a key intermediate in the biosynthesis of the pyrimidine nucleotides. The biosynthesis of orotic acid appears relatively simple, as the entire structure of aspartic acid is utilized. The pathway of orotic acid biosynthesis and its conversion into UMP are illustrated in Figure 20.

Orotic acid is converted into 5′-UMP, and the other pyrimidines are derived from UMP (Figure 21).

NUCLEOTIDE ISOLATION

Nucleotides are extracted from tissues by perchloric acid or trichloroacetic acid. The insoluble salts of nucleotides provided the first basis for nucleotide purification. Silver precipitates purines under neutral or slightly acid conditions, whereas pyrimidines are precipitated in basic solutions. Barium precipitates most nucleotides and other phosphate-containing compounds. However, it was not until 1950 that a refined method was available for nucleotide purification. In that year Cohn (8) introduced the use of ion exchange resins for the separation of nucleotides (hydrolyzed RNA). Later Hurlbert *et al.* (9) used the Dowex resins to separate the many acid-soluble nucleotides.

Formic acid–ammonium formate, HCl-$CaCl_2$, or HCl-NaCl elution solvent systems are employed to fractionate the nucleotides on resins.

A typical Dowex-1 chromatogram of nucleotides from corn tissue is shown in Figure 22.

The adsorption and elution of nucleotides from ion exchange columns actually involve chemical reactions. The reaction of a nucleotide with a strongly basic anion exchange resin such as Dowex-1, 2, 11, or 21 is illustrated in Figure 23 and the following reactions of adsorption and elution:

(1) Adsorption: $2\,(RNR_3{}^1) + 2\,HCOOH + \underset{\text{Nucleotide}}{Na_2RPO_4} \rightleftharpoons (RNR_3{}^1)_2\,RPO_4{}^{2+} + 2\,NaCOOH$

(2) Elution: $(RNR_3{}^1)_2RPO_4{}^{2+} + 2\,NH_4COOH \rightleftharpoons 2\,(RNR_3{}^1)COOH + \underset{\text{Nucleotide}}{(NH_4)_2PRO_4}$

FIGURE 20. The pathway of orotic acid synthesis.

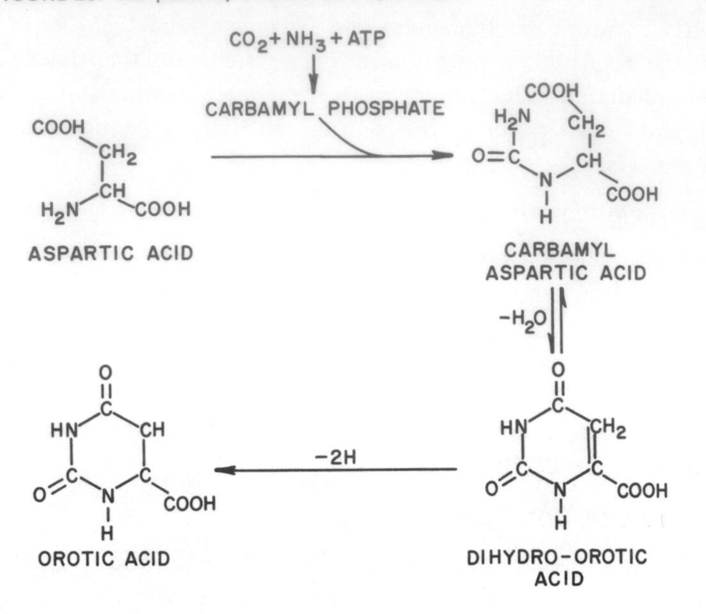

FIGURE 21. The pathway of uridylic acid (UMP) synthesis.

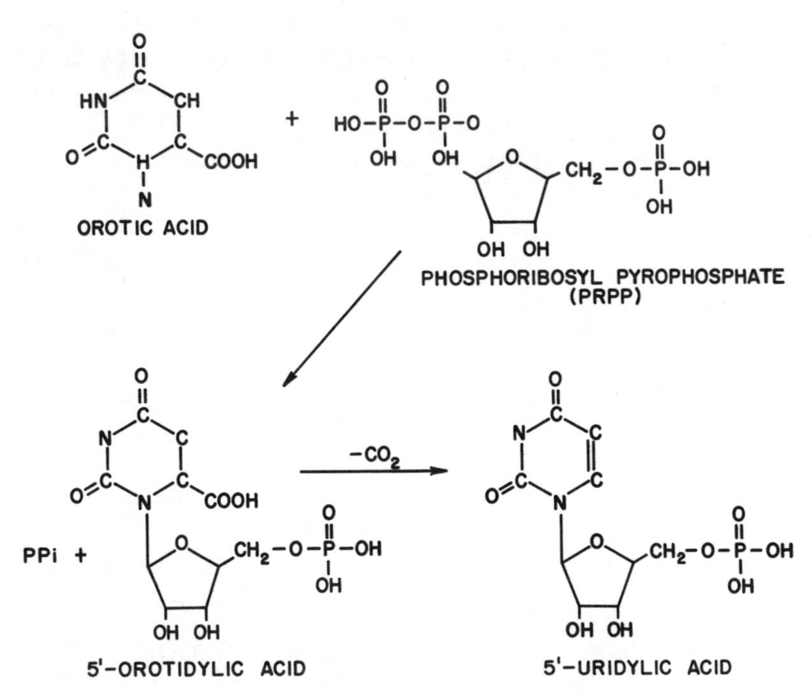

FIGURE 22. Fractionation of the acid-soluble nucleotides from corn on a Dowex-1 column (10).

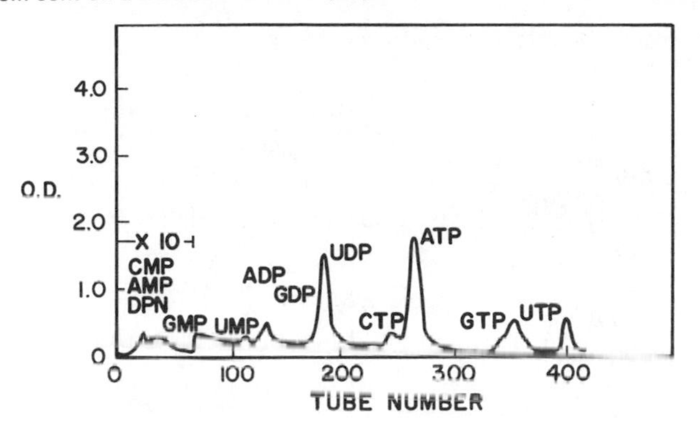

FIGURE 23. Resin group (polystrene backbone). The functional group of an anionic resin.

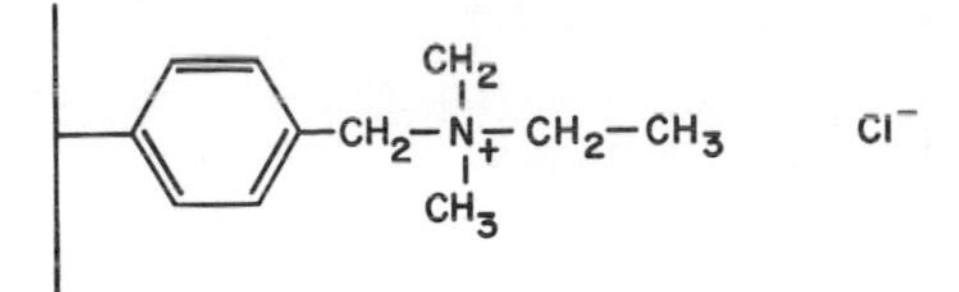

The purine or pyrimidine bases of nucleotides have unique ultraviolet absorption spectra. Therefore the U.V. absorption characteristics have been used with success in identifying nucleotide bases. See the spectra of CMP in Figure 24 and that of GMP in Figure 25. For more information on differences among nucleotide spectra see the publication of Markham (11).

Generally, nucleotides are identified by using paper chromatography, paper electrophoresis, or thin-layer chromatography. A number of solvent systems have been used for chromatography and electrophoresis.

THEORETICAL CONSIDERATION OF GRADIENTS

Often when students begin work on gradient elution procedures, a great deal of confusion arises regarding the type of gradient to be used. The gradient maker to be considered in this discussion is composed of two cylinders and is connected by a small tube (Figure 26). The mixing chamber, 1, is stirred continuously by a magnetic stirrer. From chamber 1 a solution flows into the chromatographic column. Let us consider the type of gradient that the gradient maker with the following properties will make. Let

C_1 = concentration in chamber 1

C_2 = concentration in chamber 2

FIGURE 24 (left). Ultraviolet absorption spectra of CMP at various pH's.

FIGURE 25 (right). Ultraviolet absorption spectra of GMP at various pH's.

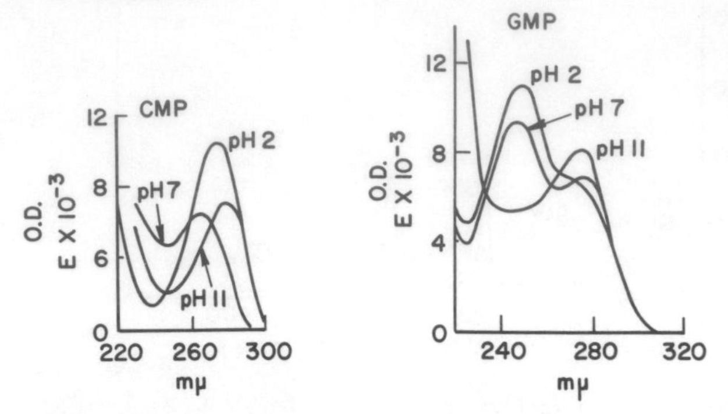

FIGURE 26. A gradient maker composed of two cylinders having the same volume (500 ml) and the same cross-sectional area.

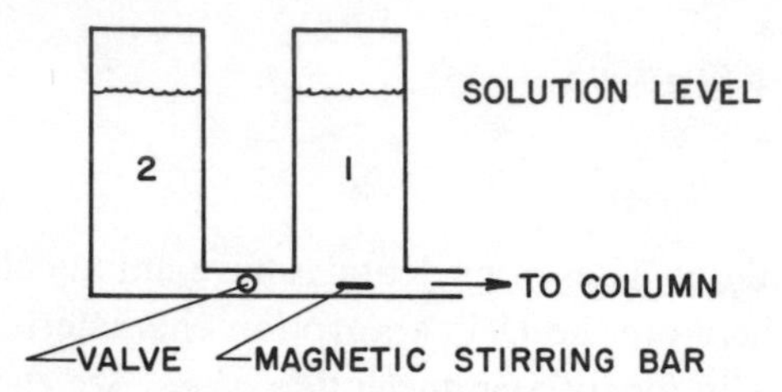

V_1 = volume in 1 V_2 = volume in 2
A_1 = cross-sectional area of 1 A_2 = cross-sectional area of 2
c = concentration of effluent coming from gradient maker
v = volume of effluent that has left the gradient maker
V = total volume of system

Then the equation to estimate the concentration of the effluent (c) is

$$c = C_2 - (C_2 - C_1)\left(1 - \frac{v}{V}\right)^{A_2/A_1}$$

Let us assume that

$V_2 = 90$ ml $C_2 = 1\ M$ NaCl
$V_1 = 10$ ml $C_1 = 0$
$v = 10$ ml $A_2/A_1 = 4$

Therefore

$$c = 1 - (1 - 0)(1 - 10/100)^4$$
$$c = 1 - (1)(0.9)^4$$
$$c = 1 - (0.66)$$
$$c = 0.34\ M\ \text{NaCl}$$

FIGURE 27. A gradient maker in which the cross-sectional areas of the two cylinders change as the liquid in the apparatus flows out.

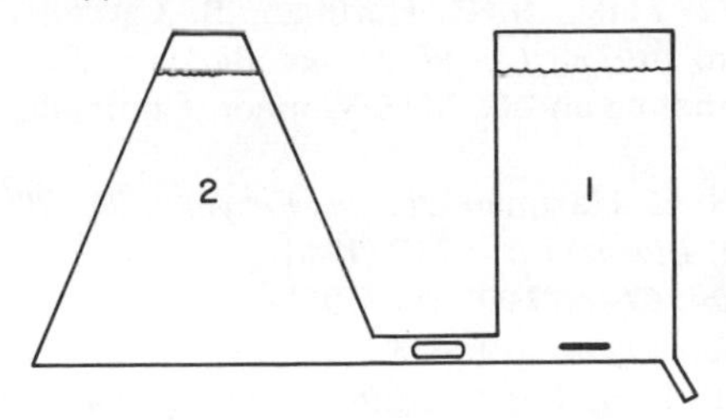

FIGURE 28. A salt gradient produced by the gradient maker illustrated in Figure 27.

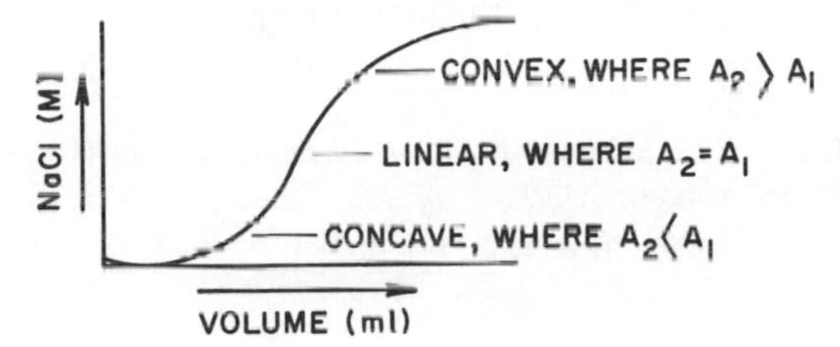

Now let us calculate the concentration of NaCl after another 10 ml passes through the system.

$$c = 1 - (1 - 0.34)(1 - 10/90)^4$$
$$c = 1 - (0.66)(8/9)^4$$
$$c = 1 - (0.66)(0.62)$$
$$c = 1 - 0.41$$
$$c = 0.59\ M\ \text{NaCl}$$

It follows from these calculations that this system (one with A_2 greater than A_1) yields a convex gradient. Thus two-cylinder gradient makers in which the volumes of both cylinders are under the same pressure will give different gradients depending on the cross-sectional areas of the cylinders. Therefore, if

$A_2 > A_1$, a convex gradient is produced
$A_2 = A_1$, a linear gradient is produced
$A_2 < A_1$, a concave gradient is produced

Now we consider a differently shaped gradient maker (see Figure 27). From the preceding discussion it is possible to predict the shape of the gradient profile (see Figure 28).

More complicated gradients can be produced, depending on the number of chambers and the shape of each of them. The general equation used in this discussion may not apply to multichamber systems.

REFERENCES

1. Buchanan, J. M., J. G. Flaks, S. C. Hartman, B. Levenberg, L. N. Lukens and L. Warren. In *The Chemistry and Biology of the Purines*. Eds. G. E. W. Wolstenholme and C. M. O'Connor. Churchill, London. 1957, p. 232.
2. Buchanan, J. M. and S. C. Hartman. *Advan. Enzymol.* 21: 200 (1959).
3. Greenberg, G. R. *Arch. Biochem.* 19: 337 (1948).
4. Greenberg, G. R. *J. Biol. Chem.* 190: 611 (1951).
5. Greenberg, G. R. *J. Biol. Chem.* 219: 423 (1956).
6. Greenberg, G. R. and E. L. Spilman. *J. Biol. Chem.* 219: 411 (1956).
7. Potter, R. van. *Nucleic Acid Outlines*. Vol. 1. Burgess Publishing Co., Minneapolis, Minn. 1960.
8. Cohn, W. E. *J. Am. Chem. Soc.* 72: 1471 (1950).
9. Hurlbert, R. B., Hans Schmitz, A. F. Brumm and R. van Potter. *J. Biol. Chem.* 209: 23 (1954).
10. Cherry, J. H. and R. H. Hageman. *Plant Physiol.* 35: 343 (1960).
11. Markham, R. In *Modern Methods of Plant Analysis*. Vol. 4. Eds. K. Paech and M. V. Tracey. Springer-Verlag, Berlin. 1955, p. 246.

EXPERIMENT 10 NUCLEOTIDE EXTRACTION, CHROMATOGRAPHY, AND IDENTIFICATION

A. OBJECTIVE

Nucleotides, directly or indirectly, touch on all phases of metabolism. Levels of certain nucleotides indicate the type of metabolism in which the cells are engaged. Therefore techniques dealing with the quantitative extraction and analysis of nucleotides are of importance to the molecular physiologist. The objectives of this experiment are to learn techniques dealing with the extraction of soluble nucleotides and subsequent identification by column chromatography, paper electrophoresis and ultraviolet absorption characteristics.

B. EQUIPMENT AND SUPPLIES

Plant tissue (pea cotyledons)
Homogenizer
0.6 *M* $HClO_4$, 0.5 *M* KOH, 4 *N* formic acid, 4 *N* formic acid–1.6 *M* ammonium formate
Refrigerated centrifuge
Dowex-1 × 8 (formate)
Fraction collector
Glass column (1.2 × 60 cm)
Spectrophotometer
Gradient maker

Paper electrophoresis apparatus and paper
Capillary tubes
Lyophilizer
Activated charcoal (Darco G-60)
Magnetic stirrer
pH Paper
U.V. mineral light
0.05 *M* Ammonium formate, pH 3.5
CCl_4
Nucleotide standards

C. EXPERIMENTAL PROCEDURE

1. Extraction of Nucleotides

Homogenize 50 g of pea cotyledons (5 days, grown in dark) in 100 ml of cold 0.6 *M* $HClO_4$ (keep cold). Centrifuge the homogenate at 20,000 × *g* for 15 min and filter the supernatant through glass wool. Then adjust the pH of the filtrate to 7.0 with 5 *N* KOH, using pH-indicator paper. Set the solution in the cold for about 30 min to allow complete precipitation of potassium perchlorate and other insoluble salts. Remove the precipitate by centrifugation at 10,000 × *g* for 15 min. Save the supernatant in the cold for nucleotide separation (freeze if time between extraction and column chromatography is longer than 1 day).

2. Separation of Nucleotides on a Dowex-1 Column

Pack a 1.2 × 60 cm column to a height of 40 cm with Dowex-1 × 8 (200–400) mesh, formate form (see Appendix 6). Wash the resin in the column with 20 ml of H_2O, and then add half of the crude nucleotide extract (equivalent to 25 g of tissue). Drain to surface, wash column with four 20-ml volumes of H_2O, and then add 10 ml of H_2O before starting the elution gradient. The nucleotides will be eluted from the Dowex-1 column with a two-step, two-reservoir gradient system. A schematic diagram of the gradient system is given in Figure 29.

This system will produce a linear gradient. For the first stage of the two-step gradient add 300 ml of water to the mixing chamber (right side) and 300 ml of 4 *N* formic acid to the reservoir (left side). Connect the tube from the port of the mixing chamber to the column, open the valve between the reservoir and the mixing chamber, and allow elution to begin. Simultaneously apply air pressure (1–2 psi) to both chambers, as indicated, so the solution levels stay at the same height. Collect consecutive fractions containing 5 ml each. After approximately 600 ml (120 fractions) passes through the column, momentarily stop the gradient to initiate the second stage. The second stage of the gradient is obtained by filling the mixing flask (first drain out the remaining liquid) with 475 ml of 4 *M* formic acid and the reservoir with 475 ml of 1.6 *M*

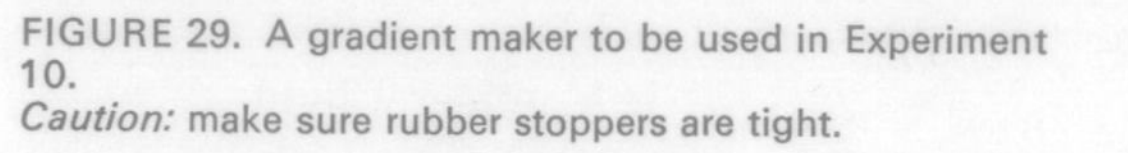

FIGURE 29. A gradient maker to be used in Experiment 10.
Caution: make sure rubber stoppers are tight.

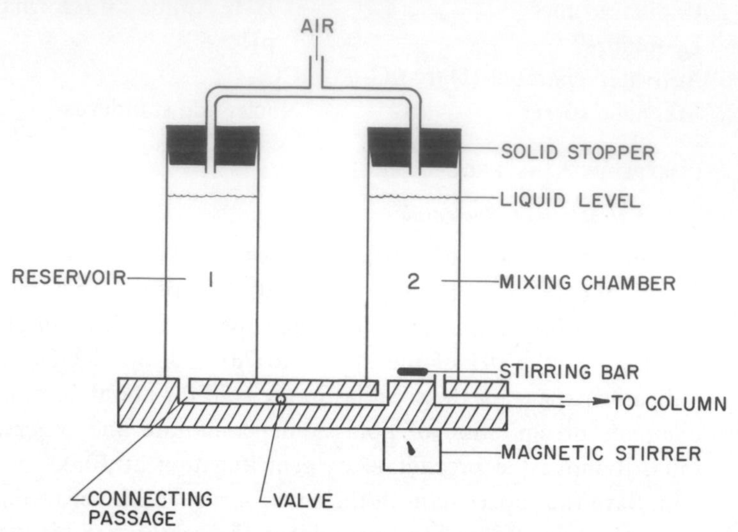

ammonium formate in 4 *N* formic acid. Again start the gradient system and collect approximately 950 ml (190 fractions). Measure the absorbancy of each fraction at 260 mμ with the spectrophotometer and plot OD_{260} against the fraction number. When finished with the Dowex column, place the used resin in a bottle marked "used Dowex-1."

3. *Purification of Nucleotide Fractions*

Combine the tubes from each of four peak fractions (your choice) and mix each fraction with 1 g of activated charcoal (Darco G-60). Stir (do not use Teflon stirring bars) the mixture for about 5 min and collect the charcoal on filter paper using a small Buchner funnel. Wash the charcoal containing the absorbed nucleotide with four 10-ml volumes of H_2O to remove salt. Carefully wash the nucleotides from the charcoal with two 10-ml volumes of 15% pyridine (work under the hood). By extraction with chloroform, remove the pyridine from this solution, which contains the nucleotides. This is best done by placing the pyridine solution in a small separatory funnel and extracting with four 15-ml volumes of chloroform. Discard the chloroform-pyridine layer and save the aqueous portion. Then lyophilize the aqueous solution to dryness. Dissolve this residue in a small amount of water (0.5–1 ml) and use for nucleotide identification. *Caution*! After using charcoal, clean all glassware with chromic acid cleaning solution.

4. Identification of Nucleotides

(a) Spectra Analysis. Use a small portion of the nucleotide samples for the determination of U.V. spectra at pH 2 and 7. Compare the spectra obtained with those for known nucleotides reported in the literature (absorbancy maximum and minimum and ratios at 250:260 and 280:260 [see (1, 2)].

(b) Paper Electrophoresis. Carefully spot the nucleotide samples (known and unknown) on Whatman No. 1 (acid-washed) paper strips as indicated in Figure 30.

(1) Spot the samples with capillary melting-point tubes and dry the moist spots with a hair dryer. Try to keep the area of the spots small (less than 0.5 cm).

(2) Soak the short end first in "running" buffer almost to the point where the samples are spotted, and then soak the long end of the sheet in the same buffer. The running buffer consists of 0.05 *M* ammonium formate, pH 3.5.

(3) Let the buffer diffuse to the sample line.

(4) Alternatively, steps (2) and (3) may be omitted, and instead the paper can be sprayed (atomizer) with enough running buffer to moisten the sheet.

(c) Procedure To Start Electrophoresis

(1) Put pieces of presoaked chromatography paper between the CCl_4 container and the running buffer containers, being careful not to get them in the CCl_4 but being sure that they are in contact with the running buffer (to provide a wick) as indicated in Figure 31.

(2) Put glass rods in the middle of the chromatograph and dip it into the CCl_4 and allow it to settle. Drape the tops of the chromatograms on the presoaked pads.

(3) Run at 500 volts, 25–50 mamp for 1 hr for nucleoside triphosphates. The di- and monophosphates should require 1.5–2 and 2–2.5 hr, respectively.

(4) After sufficient time take the paper sheet out of CCl_4 and dry it in the hood for 30 min and then in the oven for 30 min (or merely overnight in the hood).

(5) With the aid of a U.V. mineral light, mark the nucleotide spots on the paper and compare electrophoretic mobility values of the unknowns with knowns.

(6) In case the samples are radioactive, divide the strip into 0.5-in. consecutively numbered pieces. Then put the pieces into scintillation vials and measure radioactivity.

FIGURE 30. Application of nucleotide samples to a paper sheet before electrophoresis.

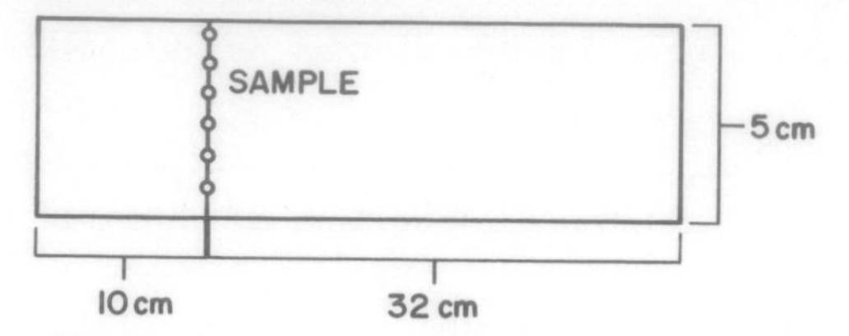

FIGURE 31. Diagram of a paper electrophoresis apparatus used for the separation of nucleotides.

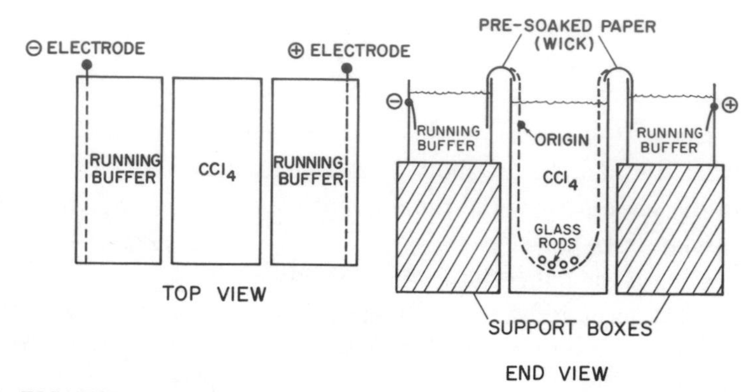

D. TREATMENT OF DATA

Summarize your data in figures, graphs, or tables and write a report in the form of a publication for *Plant Physiology.*

REFERENCES

1. Markham, R. In *Modern Methods of Plant Analysis*. Vol. 4. Eds. K. Paech and M. V. Tracey. Springer-Verlag, Berlin. 1955, p. 246.
2. Pabst Laboratories Circular OR-10 (1956).

General References

Block, R. J., E. L. Durrum and G. Zweig. *A Manual of Paper Chromatography and Paper Electrophoresis*. Academic Press, New York. 1961, p. 710.

Cherry, J. H. and R. H. Hageman. *Plant Physiol.* 35: 343 (1960).

Fink, K. and W. S. Adams. *J. Chromatog.* 22: 118 (1966).

Hulbert, R. B., H. Schmitz, A. F. Brumn and R. van Potter. *J. Biol. Chem.* 209: 23 (1954).

Ingle, J. *Biochim. Biophys. Acta* 61: 142 (1962).

Keys, J. K. *J. Exp. Bot.* 14: 16 (1963).

Robern, H., D. Wang and E. R. Waygood. *Can. J. Biochem.* 42: 345 (1965).

Robern, H., D. Wang and E. R. Waygood. *Can. J. Biochem.* 43: 225 (1965).

Wilson, A. T. and D. J. Speddling. *J. Chromatog.* 18: 76 (1965).

PART

6

PROPERTIES OF NUCLEIC ACIDS

INTRODUCTION

The story of nucleic acids begins with a student named Miescher who in 1869, at the age of 25, set out to characterize cell nuclei. Miescher digested pus cells with pepsin in the presence of hydrochloric acid. Nuclei remained in the aqueous phase after ether extraction and settled out with time. A material called "nuclein" isolated from nuclei was acidic in nature, insoluble in dilute acids but soluble in dilute alkali. Furthermore, the material contained a considerable amount of phosphorus. The discovery of this new material, nucleic acid, was not

FIGURE 32. Structure of an oligoribonucleotide (a short piece of RNA).

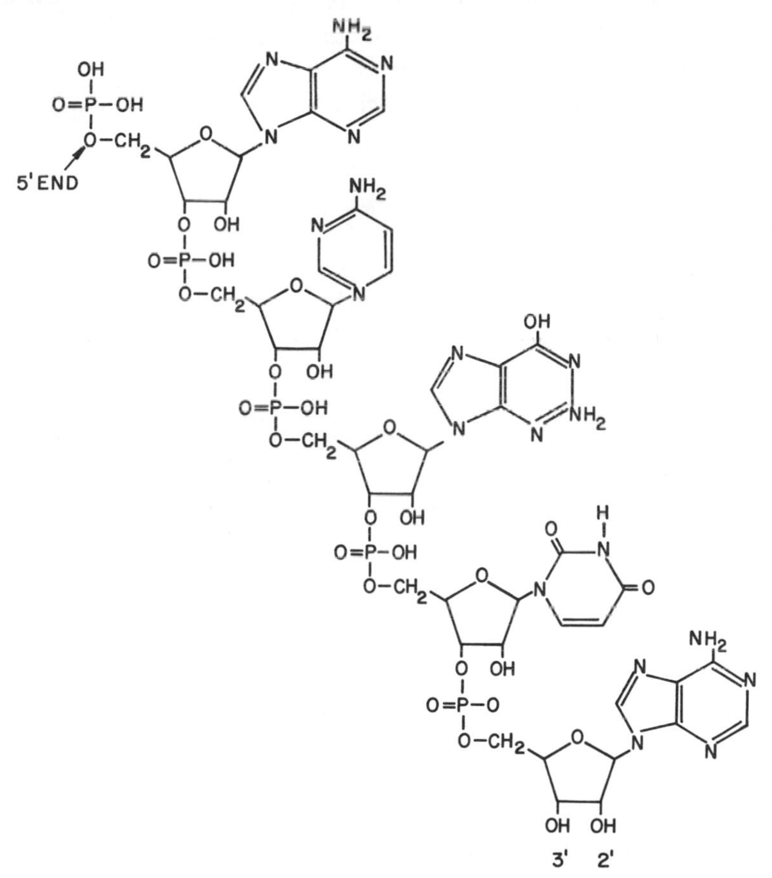

accepted by Miescher's major professor, Hoppe-Seyler, until he had personally repeated the experiments. When both Miescher's and Hoppe-Seyler's papers were published, a whole new area of molecular biology was initiated. For many years after the discovery, most experiments dealt with the chemical characterization of nucleic acids. However, the modern era of nucleic acid chemistry did not come into being until 1950.

STRUCTURE

The structures of nucleic acids consist of long sequences of nucleosides (several hundred to thousands) linked together by phosphates in diester bonds from the 3′ hydroxyl of one pentose to the 5′ hydroxyl of the next pentose (Figure 32). In ribonucleic acids (RNA) the pentose is D-ribose, while it is 2-D-deoxyribose in the deoxyribonucleic acids (DNA). RNA contains the four bases adenine, guanine, cytosine, and uracil; DNA contains adenine, guanine, cytosine, and thymine (see Figure 16 in Part 5 for structures of the bases).

The complicated form in Figure 32 can be greatly simplified as

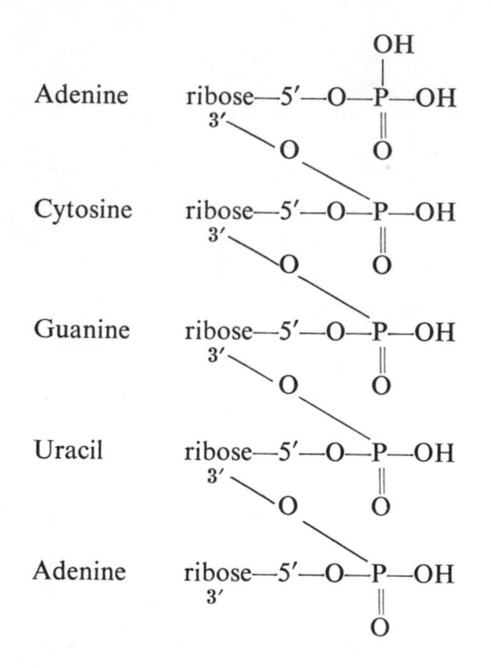

Since the pentose (ribose or deoxyribose) and phosphate backbone remain the same with only the bases varying, even this abbreviated

form can be further shortened to

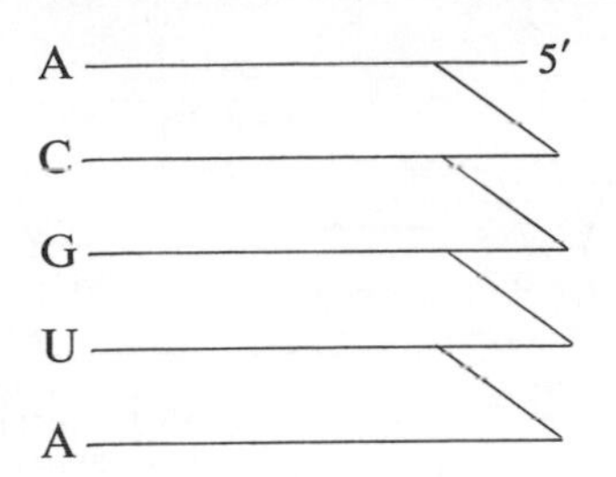

CHEMICAL PROPERTIES

1. General

Nucleic acids generally act in solutions like polyphosphoric acid (strong polyanions) and thus are usually soluble in water as salts and as acids. Nucleic acids are, however, usually insoluble in nonpolar solvents. The solubility characteristics of nucleic acids provide the fundamental basis of most purification procedures.

2. Alkali Hydrolysis

Much attention has been given to the action of dilute alkali (0.1–1.0 *N*) on nucleic acids at moderate temperatures. RNA is completely hydrolyzed to a mixture of 2′- and 3′-nucleoside monophosphates. However, under the same conditions DNA is not attacked but is merely solubilized. The initial action of alkali on RNA is to convert phosphate diester linkages into a cyclic 2′,3′ intermediate, which is then hydrolyzed to a mixture of the 2′ and 3′ monoesters (see Figure 33). Since RNA usually contains a 3′-OH terminal, alkali will remove the 5′-phosphate linkage and yield a nucleoside. Thus, for every molecule of RNA, alkali hydrolysis will produce one nucleoside. At the other end of the RNA molecule, 5′-phosphate, alkali yields a 2′(3′),5′-nucleoside diphosphate.

DNA is resistant to alkali hydrolysis owing to the absence of the 2′-hydroxyl on deoxyribose. This chemical difference in RNA and DNA was employed in the method to measure selectively the contents of the two nucleic acids (Schmidt-Thannhauser).

3. Acid Hydrolysis

Acid hydrolyzes both RNA and DNA to a wide array of different materials containing 3′- and 5′-nucleoside phosphates, nucleosides, and free bases. The purine-pentose linkage is broken in acid more readily than that of the pyrimidine-pentose. Therefore heating nucleic acids at 100 C for 10 min removes the purines, whereas 1 hr is required to remove the pyrimidines.

FIGURE 33. Hydrolysis of RNA by alkali.

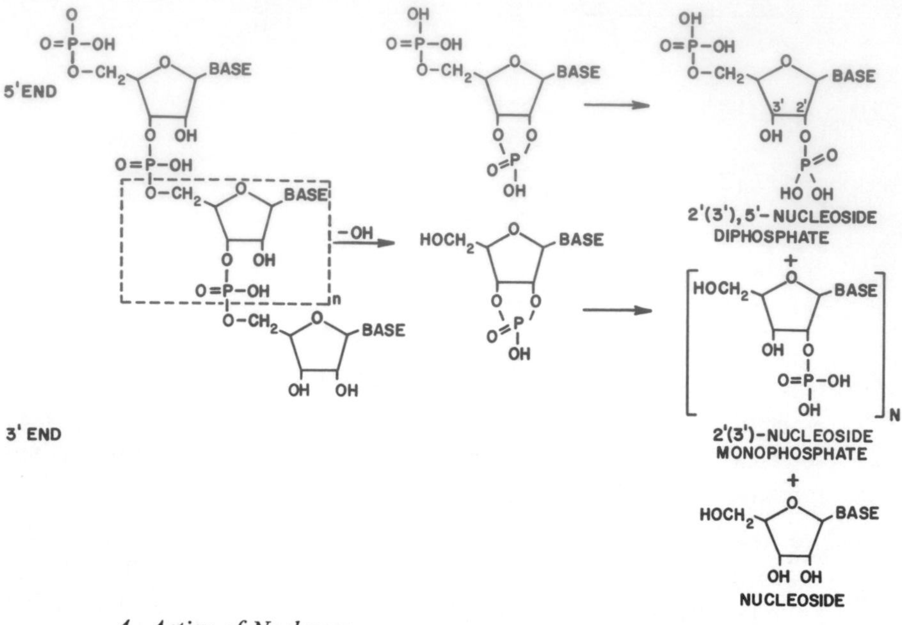

4. Action of Nucleases

A number of enzymes that hydrolyze nucleic acids to the nucleoside monophosphates have been isolated from several sources and characterized. The specificities of the enzymes involve the following characters of the substrate:

(a) Site of attack (at the end of the molecule or at random).

(b) Presence of phosphate or hydroxyl end groups.

(c) Presence of purines or pyrimidines around the phosphate diester.

(d) Preference for attacking the C-3′-P or the C-5′-P linkage.

(e) Preference for nucleic acids containing ribose or deoxyribose.

The specificity of many of the nucleases has become extremely valuable in the determination of nucleotide sequences. However, of the various nucleases studied, the mechanism of action is best known for pancreatic RNase. This enzyme is an endonuclease that splits 5′-phosphate links adjacent to 3′-pyrimidines at random within the RNA molecule. Pancreatic RNase attacks RNA until the maximum number of 2′(3′)-pyrimidine ends exist. The enzyme requires that the 2′-hydroxyl be available because a 2′,3′ cyclic phosphate intermediate is involved.

Other nucleases, notably snake venom RNase, act as exonucleases and attack RNA from the 3′-phosphate linkage to yield 5′-nucleoside monophosphates. Many other specific examples of nuclease action are known but they are not summarized here.

ISOLATION

Methods of nucleic acid isolation are based on insolubility or solubility as acids or salts in polar or nonpolar solvents. The degree of purity in the RNA isolated is dependent upon the source of material. That is, some plant material contains a higher concentration of contaminants (DNA) than others. Fractionation of cell homogenates into subcellular fractions could be of importance in isolating different types of nucleic acids. For example, the use of nuclei could be of assistance in obtaining purified DNA. The cytoplasmic fraction could be the source of soluble and ribosomal RNAs. A wide array of other techniques involving column fractionation is usually employed in nucleic acid purification, which is discussed in Parts 8 and 9.

In essence, most isolation methods involve these characteristics of nucleic acids:

1. Insolubility in acids.
2. Insolubility in various organic solvents (alcohol, ether, acetone, etc.).
3. Insolubility of heavy metal salts of nucleic acids in aqueous solutions.
4. Unique ultraviolet absorption spectrum of nucleic acids.
5. Property of RNA to be hydrolyzed in alkali while DNA is not.
6. Specific reaction of RNA and DNA with orcinol and diphenylamine, respectively.

The Schmidt-Thannhauser method was one of the first procedures to utilize the unique solubility characteristics of nucleic acids that have been modified in several aspects. Two of these methods are employed in Experiment 11. Other methods devised to obtain purified RNA or DNA are presented in Parts 8 and 9.

GENERAL REFERENCES

Chargaff, E. and J. N. Davidson, eds. *The Nucleic Acids*. Vol. 1. Academic Press, New York. 1955.

Chargaff, E. and J. N. Davidson, eds. *The Nucleic Acids*. Vol. II. Academic Press, New York. 1955.

Chargaff, E. and J. N. Davidson, eds. *The Nucleic Acids*. Vol. III. Academic Press, New York. 1960.

Markham, R. In *Modern Methods of Plant Analysis*. Vol. 4. Eds. K. Paech and M. V. Tracey. Springer-Verlag, Berlin. 1955, p. 246.

Potter, R. van. *Nucleic Acid Outlines*. Vol. 1. Burgess Publishing Co., Minneapolis, Minn. 1960.
Schmidt, G. and S. J. Thannhauser. *J. Biol. Chem.* 161: 83 (1945).

EXPERIMENT 11 MEASUREMENT OF NUCLEIC ACIDS IN PLANTS

A. OBJECTIVE

Changes in nucleic acid content often reflect important biological events in many organisms. Therefore a knowledge of the procedures used to measure RNA and DNA is valuable to most students. The objective of this experiment is to learn techniques dealing with the isolation and measurement of nucleic acids.

B. EQUIPMENT AND SUPPLIES

Peanut seedlings (cotyledons and roots)
Tomato leaves
Cold methanol
Cold 0.2 *M* $HClO_4$
Cold ethanol (95%)
Ethanol: ether (2:1)
0.5 *M* KOH
5% $HClO_4$
Reagents for diphenylamine test
Refrigerated centrifuge
Glass homogenizer with power-driven Teflon pestle
Water bath
Spectrophotometer

C. EXPERIMENTAL PROCEDURE

1. General Procedure

Tissues from cotyledons and root tips (1 cm) of 2-day-old peanut seedlings and leaves of green tomato will be assayed for RNA and DNA contents (see the flow sheet below). In quadruplicate, homogenize 0.5-g samples in 2 ml of cold methanol with an ice-jacketed glass homogenizer and a power-driven Teflon pestle (see Experiment 5). Transfer the homogenate to polypropylene centrifuge tube, and then wash the glass homogenizer twice with 2 ml of methanol. Centrifuge the combined homogenate and two washes at 5000 × *g* for 10–15 min and discard the supernatant. Wash each pellet twice with 4-ml volumes of cold methanol and discard supernatants. Extract the methanol-insoluble material twice with 4 ml of cold 0.2 *M* $HClO_4$ to remove acid-soluble phosphates (keep the samples cold to prevent hydrolysis of nucleic acid). The acid precipitate is next washed (suspend and then centrifuge) twice with 4 ml of cold ethanol to remove the acid and other soluble materials. This step is followed by extraction of the lipids with ethanol:ether (2:1) at 50 C for 30 min. Centrifuge and discard the supernatant. At this point, the precipitate contains mainly nucleic acid, protein, starch, and cellulose. RNA can be removed by KOH hydrolysis (see below). However, both RNA and DNA are extracted (hydrolyzed) in hot $HClO_4$. Therefore, to compare these two methods of nucleic acid extraction, two samples of each set will be carried through the two following procedures.

2. Method A

To one set of the tubes add 5 ml of 0.5 *M* KOH and incubate for 16 hr at 37 C. After incubation, centrifuge and collect the supernatant, which contains the hydrolyzed RNA [2′(3′)-nucleoside monophosphates]. Wash the pellet once (centrifuge at 5000 × *g* for 10–15 min to set the pellet) with 1 ml of 0.5 *M* KOH and add the wash to the first supernatant. The DNA is then extracted from the pellet (or KOH-insoluble material) by incubation with 5 ml of 5% $HClO_4$ at 70 C for 40 min. After this incubation step, centrifuge and collect the supernatant, which contains hydrolyzed DNA. Wash with 1 ml of 5% $HClO_4$ and combine with first supernatant. For both the KOH and $HClO_4$ hydrolysates, adjust the pH to approximately 7, adjust the volume to 10 ml with H_2O, and then centrifuge in the cold to remove $KClO_4$. Read the supernatant at 260 and 290 mμ on the spectrophotometer. Multiply the 260 − 290 reading by 57 to obtain micrograms of nucleic acid per milliliter. For one sample of each tissue, determine the U.V. absorption spectrum from 220 to 300 mμ by taking readings at 5-mμ intervals.

3. Method B

To the other set of tubes add 5 ml of 5% $HClO_4$ and incubate at 70 C for 40 min. After incubation, centrifuge and collect the supernatant. Wash by centrifugation the insoluble pellet once with 1 ml of 5% $HClO_4$ and add this supernatant to the first. Adjust the pH to approximately 7, adjust the volume to 10 ml with H_2O, and then collect by centrifugation in the cold the insoluble $KClO_4$. Determine the U.V. absorbancy at 260 and 290 mμ of each supernatant and an absorption spectrum (between 220 and 300 mμ) for one sample of each tissue. Again the factor of 57 is used to calculate total nucleic acids.

DNA content is determined by employing the diphenylamine test for deoxyribose. Incubate 1 ml of the neutralized acid extract with 2 ml

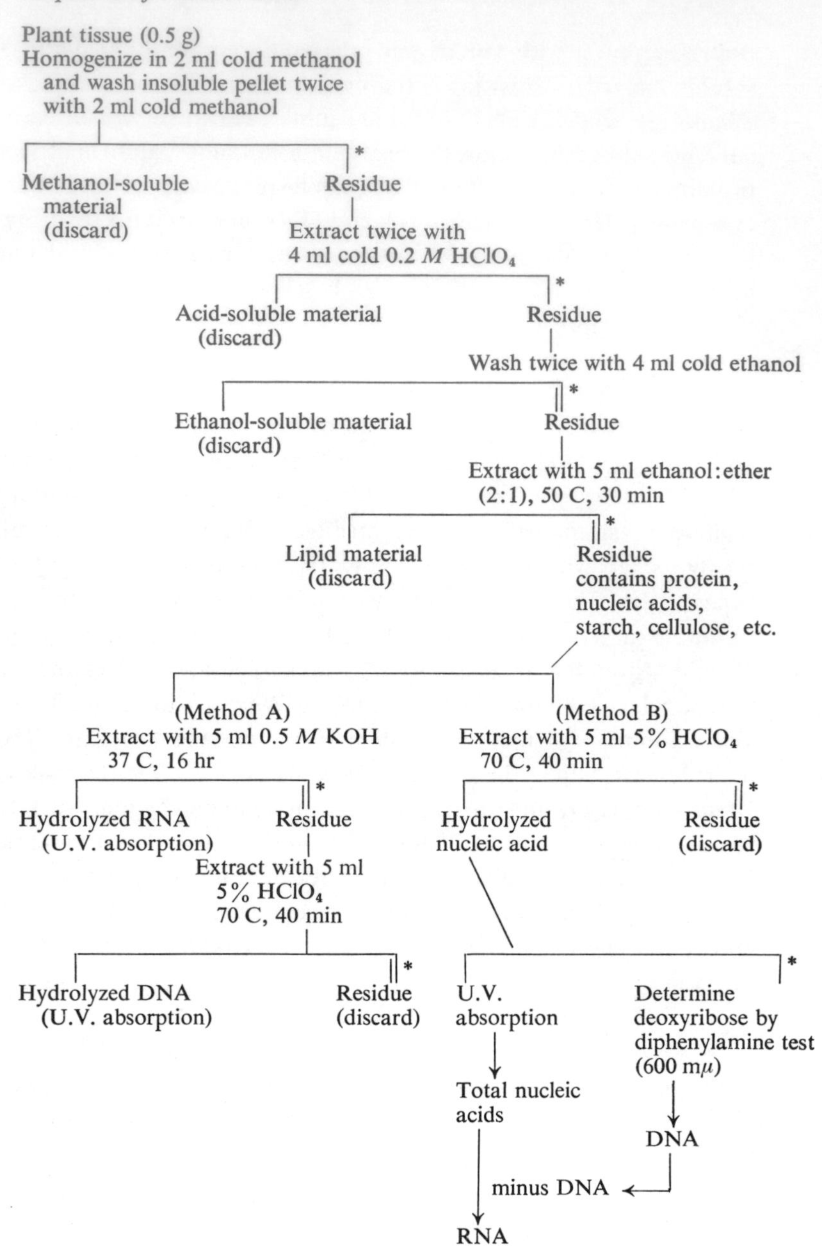

* = Involves centrifugation at 5000 × g for 10–15 min. Use polypropylene tubes.

of diphenylamine reagent* at room temperature for 16 hr in the dark. As a standard, prepare tubes which contain 25–100 μg of DNA in 5% $HClO_4$ and a blank containing only 5% $HClO_4$. Determine the absorbancy at 600 mμ. When finished with the experiment, wash the tubes with alcohol.

Subtract the amount of DNA estimated by the diphenylamine test from total nucleic acids to obtain the amount of RNA.

D. TREATMENT OF DATA

Summarize your data in figures, graphs, or tables and write a brief report of your results.

GENERAL REFERENCES

Burton, K. *Biochem. J.* 62: 315 (1956).
Cherry, J. H. *Plant Physiol.* 37: 670 (1962).
Markham, R. In *Modern Methods of Plant Analysis.* Vol. 4. Eds. K. Paech and M. V. Tracey. Springer-Verlag, Berlin. 1955, p. 246.
Ogue, M. and G. Rosen. *Arch. Biochem.* 25: 262 (1950).
Potter, R. van. *Nucleic Acid Outlines.* Vol. 1. Burgess Publishing Co., Minneapolis, Minn. 1960.
Schmidt, G. and S. J. Thannhauser. *J. Biol. Chem.* 161: 83 (1945).
Schneider, W. *J. Biol. Chem.* 161: 293 (1945).

EXPERIMENT 12 ALKALI HYDROLYSIS OF RNA AND FRACTIONATION OF NUCLEOTIDES

A. OBJECTIVE

Alkali hydrolyzes the phosphate diester links of RNA to yield the 2′(3′)-nucleoside monophosphates. The objective of this experiment is to learn the techniques of fractionating the nucleoside monophosphates of RNA on a Dowex-1 column.

* The diphenylamine reagent is prepared by dissolving 1.5 g of diphenylamine in 100 ml of redistilled acetic acid (or by using a new bottle of glacial acetic acid) and adding 1.5 ml of concentrated H_2SO_4. Store the diphenylamine reagent in the dark. On the day the reagent is to be used, add 0.1 ml of aqueous acetaldehyde (16 mg/ml) for each 20 ml of reagent required.

B. SUPPLIES AND EQUIPMENT

0.5 *M* KOH
5 *M* $HClO_4$
Formic acid solutions (1.5 *N*, 4 *N*, and 6 *N*)
Water bath
Spectrophotometer
Yeast RNA
Dowex-1 × 8 (formate)
Glass column, 1.2 × 50 cm
Fraction collector

C. EXPERIMENTAL PROCEDURE

1. Alkali Hydrolysis

To a solution of 40 OD units (2 mg) of yeast RNA dissolved in a few milliliters of H_2O add enough 4 *N* KOH to make the final concentration 0.5 *N*. Incubate for 16 hr at 37 C to hydrolyze the RNA. Then adjust the pH of the solution to 7.0 with 5 *M* $HClO_4$ and centrifuge in the cold for 30 min to collect the precipitate. Save the supernatant for chromatographic separation of the nucleotides.

2. Dowex-1 Column

Prepare a Dowex-1 × 8 (formate)* column in a glass tube (1 cm × 50 cm) to a height of 15 cm. Dilute the hydrolyzed RNA sample (from C-1) with 20 ml H_2O and add to the column (make sure the pH is near 7.0). Then wash the column with about 20 ml H_2O.

The nucleotides will be separated on the Dowex column with a three-step gradient of formic acid. The formic acid reservoir and the mixing chamber of the gradient apparatus used in Experiment 10 will be used to establish the gradient. However, in this experiment a constant volume in the mixing chamber will be maintained. To do this, close the valve between the two chambers. Then add 500 ml of H_2O to the mixing chamber and 250 ml of 1.5 *N* HCOOH to the reservoir and start the stirrer. Apply about 2–3 psi of air pressure to both chambers and then clamp off the air line to the mixing chamber. Open the valve and allow elution to begin. Collect the eluate in 6-ml fractions, using a fraction collector. When the first 250 ml has passed through the system, close the valves and add 500 ml of 4.0 *N* HCOOH to the reservoir. Again apply air pressure as above. Open the valves and begin the elution process. Finally, when the 4 *N* HCOOH has passed through the system, momentarily stop the elution and add 250 ml of 6.0 *N* HCOOH to the reservoir and continue elution in the manner described above. Read the U.V. absorbance at 260 mμ and plot the optical density (OD) against fraction number. With this system, the mononucleotides of RNA should be fractionated in a manner illustrated in Figure 34.

* See Appendix 6 for the procedure to prepare Dowex.

FIGURE 34. Fractionation of the ribonucleoside monophosphates (alkali hydrolysis of RNA) on a Dowex-1 column.

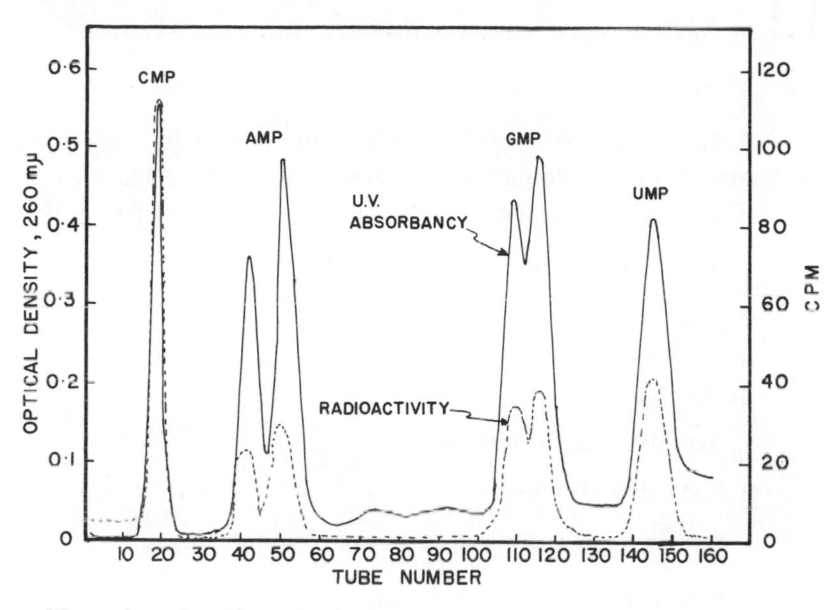

Note that the 2′- and 3′-phosphates of adenosine and guanosine are separated by this procedure.

3. Calculation of Results

Sum the absorbancy of each peak fraction and subtract the base line. Using the molar absorbancy values at pH 2 [see (1)] of 15.1 × 10³ for AMP, 13 × 10³ for CMP, 10.0 × 10³ for UMP, and 12.2 × 10³ for GMP, and the equation* $a_m = A/C$, calculate the molar ratios of the four nucleotides of the RNA.

D. TREATMENT OF DATA

Illustrate your data in figures or tables and write a brief report.

REFERENCE

1. Pabst Laboratories Circular OR-10 (1956).

General References

Cherry, J. H. *Science* 146: 1066 (1964).

Markham, R. In *Modern Methods of Plant Analysis*. Vol. 4. Eds. K. Paech and M. V. Tracey. Springer-Verlag, Berlin. 1955, p. 246.

* a_m = molar absorbancy index; A = total absorbancy (OD obtained in 1 cm cell × total volume); C = concentration in moles.

EXPERIMENT

13 A SIMPLE WAY TO FRACTIONATE RNA NUCLEOTIDES

A. OBJECTIVE

This experiment provides an alternative and simpler way to fractionate the nucleoside monophosphates of RNA on a Dowex-1 column. The objectives are essentially the same as those of Experiment 12.

B. SUPPLIES AND EQUIPMENT

Same as for Experiment 12 except for these solutions: 0.1 *N* formic acid, 1 *N* formic acid, 3 *N* formic acid, and 4 *N* formic acid–0.1 *M* ammonium formate.

C. EXPERIMENTAL PROCEDURE

1. Alkali Hydrolysis

Same as for Experiment 12.

2. Dowex-1 Column

Prepare a Dowex-1 column as described in Experiment 12. Add the hydrolyzed RNA sample in 20 ml of water to the column. Instead of a three-step gradient system, as used previously, a four-step discontinuous gradient system will be employed in this experiment. Diagrammatically this system is shown in Figure 35.

The stepwise elution of the nucleotides is initiated by adding 100 ml of 0.1 *N* HCOOH to the reservoir, opening the valve, and allowing acid to flow through the column. The second, third, and fourth steps are effected by adding at the completion of each subsequent step 100 ml of 1 *N* HCOOH, 120 ml of 3 *N* HCOOH, and 100 ml of 4 *N* HCOOH containing 0.1 *M* NH_4COOH, respectively. Collect the eluate in 6-ml fractions. Determine the U.V. absorbancy at 260 mμ and plot the absorbancy (OD) against the fraction number. When this system is used, the nucleoside monophosphates should fractionate in a manner illustrated in Figure 36.

As noted from this elution pattern, the 2′,3′-phosphates of adenosine and guanosine are not separated as previously seen (Experiment 12).

3. Calculation of Results

Same as for Experiment 12.

FIGURE 35. A gradient and column system to separate nucleoside monophosphates.

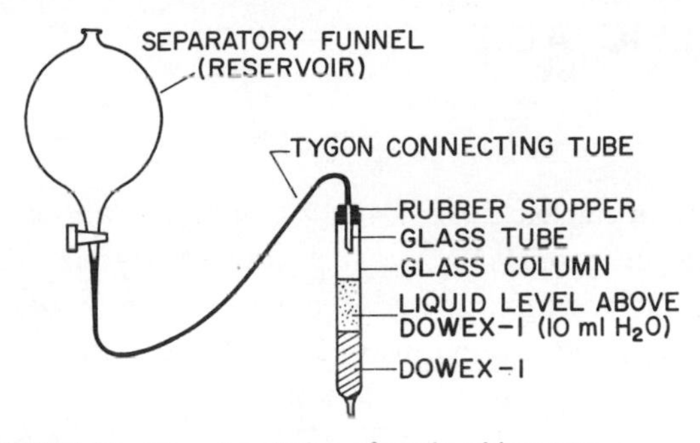

FIGURE 36. Stepwise elution of nucleoside monophosphates from a Dowex-1 column.

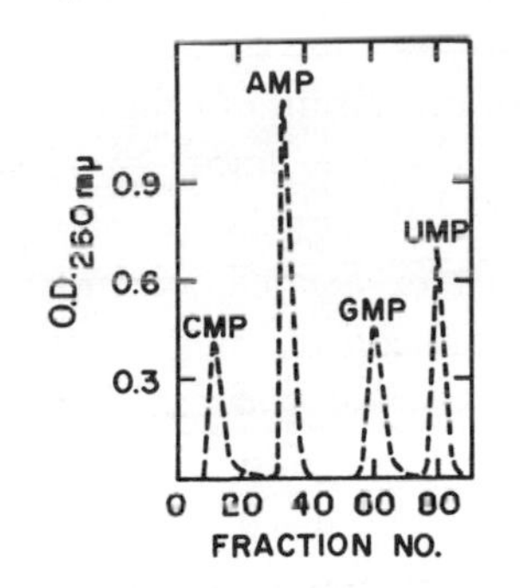

D. TREATMENT OF DATA

Illustrate your data in figures or tables and write a brief report.

GENERAL REFERENCE

Ewing, E. E. and J. H. Cherry. *Phytochemistry* 6: 1319 (1967).

PART
7
DEOXYRIBONUCLEIC ACID

INTRODUCTION

Genes are made of nucleic acid, usually DNA, and are found both in the nucleus and in cytoplasmic organelles of eucaryotic organisms. The evidence that DNA carries the genetic information was discovered in the early 1940s. It was previously demonstrated that genes are carried on cytologically distinguishable bodies (chromosomes) within the nucleus. It was then found that the DNA content per chromosome is constant. Later, a group of scientists showed that certain hereditary traits of pneumocci can be transmitted from one strain of the bacteria to another by transferring the DNA of one strain of cells to the other. These "transformation" experiments made a large impression on geneticists because the data provided a solution to the nature of the gene. However, to biochemists, the data created a wide array of questions concerning the structure, replication, and mechanism by which DNA encodes and controls genetic information. Today, after thirty years, the answers to many of these questions are available. Probably the largest impact of any scientific discovery on DNA and molecular genetics was made by Watson and Crick. Influenced by Pauling's α-helix of proteins, Watson and Crick began to test various possible models of DNA structures against the results of X-ray diffraction studies made by another research team in England. Their solution to the problem was a complementary double helix (Figure 37).

The Watson and Crick model accounted for both the X-ray diffraction data and the findings of Chargaff that the amounts of adenine and thymine are always equal and that likewise the amounts of guanine and cytosine are equal. The configuration of adenine and thymine are such that the bases form two hydrogen bonds with each other, while guanine and cytosine interact to form three hydrogen bonds. The numerous hydrogen bonds along the two complementary chains of DNA produce the force that holds the macromolecule together. The nature of these hydrogen bonds is illustrated in Figure 38.

Today there are still many biochemical problems relating to the replication and transcription of DNA. Solutions to these problems are often limited by technical difficulties. However, a few techniques dealing with the purification, characterization, and replication of DNA are discussed here.

FIGURE 37 (left). Diagram of native (double-stranded) DNA.

FIGURE 38 (right). The hydrogen bonds produced by the interaction of thymine with adenine and by cytosine with guanosine.

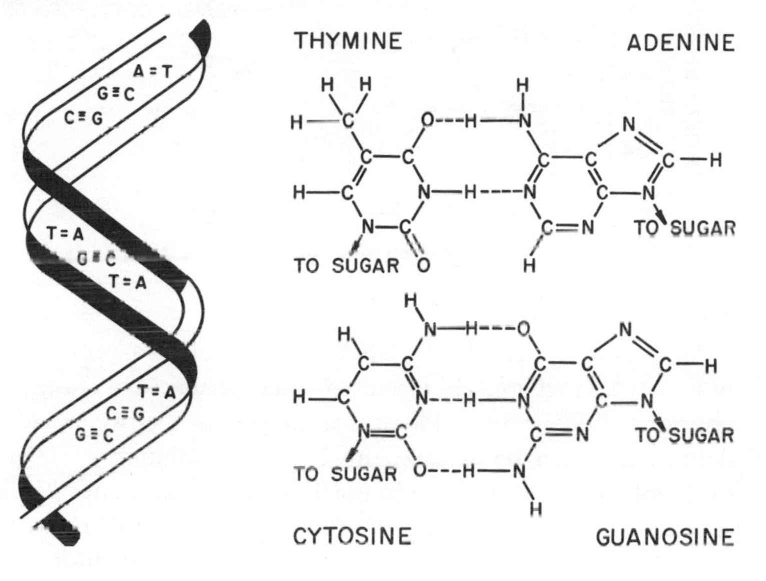

DNA ISOLATION

Several procedures for the isolation of DNA from plant materials have been developed, mostly through adaptations of the original method of Marmur (1). Generally, there are two main problems relating to DNA isolation: the isolation of the DNA-rich subcellular fraction, the nucleus, or chromatin; and the removal of the associated proteins and other macromolecular contaminants. The isolation of nuclear or chromatin material from most tissues is a fairly easy task. However, in storage tissues of plants, the presence of starch grains and protein bodies hinders the isolation of nuclei. Once a DNA-rich cellular fraction is obtained, the histones and other proteins can be dissociated from the DNA by a detergent, such as sodium lauryl sulfate in the presence of NaCl. (Frequently other agents are used to separate DNA from its chromosomal proteins and RNA. High concentrations of urea, CsCl, and $LiCl_2$ are used most often.) The dissociated proteins are removed by shaking the DNA-rich fraction with chloroform–isoamyl alcohol. They precipitate at the interphase after the mixture is centrifuged. The DNA remains soluble in the aqueous phase and can be isolated as a fibrous material on a stirring rod by gently mixing the solution with cold ethanol. Various modifications of this method are illustrated in subsequent experiments.

FIGURE 39. Melting profiles of pea DNA and peanut DNA.

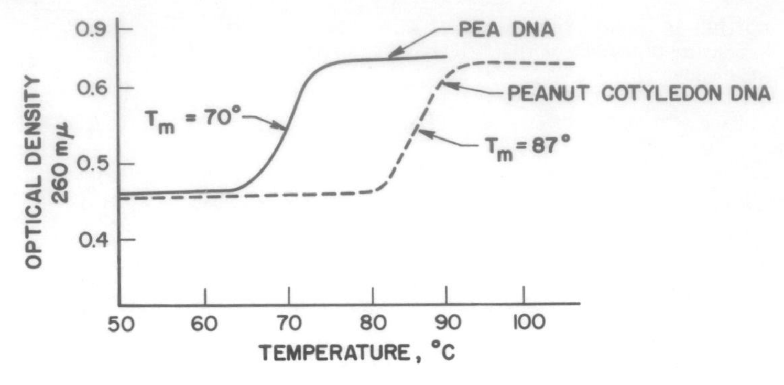

PROPERTIES OF DNA

As previously discussed, DNA consists of deoxyribonucleosides of four types that are held together by phosphate esters at the 3′ and 5′ positions of deoxyribose. Physical separation of the two strands of the double helix can be achieved by heating a solution of DNA to a sufficiently high temperature to break the hydrogen bonds holding the strands together. The process of gradually heating a DNA solution to determine the temperature range where strand separation occurs is called melting. When the DNA strands are in the form of a duplex, the highly organized structure decreases the U.V. absorbance by the bases in the individual strands. Therefore melting native DNA substantially increases (by about 40%) the U.V. absorbance of the solution. Typical melting profiles of DNA are illustrated in Figure 39.

It is obvious from Figure 39 that the two DNA samples are different. Since the difference observed by this technique is a function of the forces which hold the two strands together, it is directly related to the proportion of A–T to G–C pairs. A very meaningful characteristic of DNA is the ratio of A + T to G + C. A list of these ratios for DNAs from several sources are presented in Table 8.

TABLE 8

Source of DNA	$\frac{A+T}{G+C}$	*Source of DNA*	$\frac{A+T}{G+C}$
E. coli	0.97	Bacteriophate T2	2.03
Yeast	1.80	Wheat germ	1.17
Scendesmus (algae)	0.57	Peanut cotyledon	0.72
Rat liver	1.44	Soybean root	0.55
Calf thymus	1.38	Soybean hypocotyl	0.98
Human spleen	1.44	Tobacco leaf	1.77
Mouse spleen	1.44	Salmon sperm	2.03

Marmur and Doty (2) showed that a linear relationship between the bases composing DNA can be expressed in terms of the percentage of guanine and cytosine bases and the denaturation temperature, T_m. To determine the base composition of DNA using this technique, the DNA is carefully melted in a solution containing 0.2 *M* Na^+. The T_m (the temperature for half the maximum denaturation) so obtained is then inserted in the equation

$$T_m = 69.3 + 0.41(G + C)$$

where

$$T_m = {}^\circ C$$

and

$$G + C = \text{mole percent of } G + C$$

Since G is equal to C, the value for $(G + C)/2$ is equal to the mole percent of G and C each. Furthermore $[100 - (G + C)]/2$ is equal to the mole percent of A and T each.

When the denaturation temperature of DNA is used to calculate base composition, note that an increase in the salt concentration raises the observed T_m value. The melting profiles in Figure 39 were obtained in a low salt concentration. However, note also that the T_m value is independent of the molecular weight of DNA.

REPLICATION OF DNA

One of the great marvels of DNA is that replication of each strand yields two double helical structures exactly like the original one. This, of course, is a necessary requirement before cell division can take place. The enzyme, DNA polymerase, that catalyzes the *in vitro* synthesis of DNA was isolated from *E. coli* by Kornberg (3). Using a DNA primer, the reaction was shown to incorporate the four deoxyribonucleoside triphosphates into DNA, with a concomitant liberation of pyrophosphate:

$$n \text{ dNTP's} \xrightarrow[\text{DNA polymerase}]{\text{DNA template}} (\text{DNA})_n + n \text{ PPi}$$

The *in vitro* reaction requires an unwinding of the native DNA while the strands are copied. The exact mechanism of *in vivo* DNA replication is not yet fully understood. For example, the Kornberg enzyme appears not to be the true "replicase"; instead, this *E. coli* DNA polymerase can be more adequately described as a "repairase" or a "recombinase."

Before cell division occurs, all the DNA of the nucleus is replicated

FIGURE 40. Transcription of native DNA by DNA polymerase.

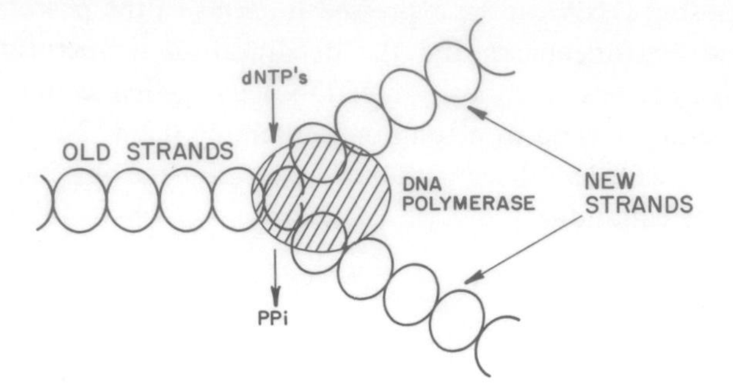

by DNA polymerase to yield a full complement of the genetic information for each of the daughter cells. It is believed that DNA replication occurs as shown in Figure 40.

DNA polymerase from plant tissues has been studied in only a few cases. The enzyme has been demonstrated in roots of mung bean and corn seedlings. Although the specific activity of the crude extracts is about 0.1% of that of the bacterial enzyme, the activity of DNA polymerase per unit weight of DNA is the same for both plants and bacteria. The enzyme also is associated with chromatin and can be assayed with the same preparations. The enzyme requires all the dNTP's, Mg^{2+}, and a sulfhydryl reducing agent for optimal activity (4).

REFERENCES

1. Marmur, J. *J. Mol. Biol.* 3: 208 (1961).
2. Marmur, J. and P. Doty. *J. Mol. Biol.* 5: 109 (1962).
3. Kornberg, A. In *Chemical Basis of Heredity*. Eds. W. McElroy and B. Glass. Johns Hopkins Press, Baltimore. 1957.
4. Leffler, H. R., T. J. O'Brien, D. V. Glover and J. H. Cherry. *Plant Physiol.* 48: 43 (1971).

General References

Bonner, J. In *Plant Biochemistry*. Vol. 7. Eds. J. Bonner and J. Varner. Academic Press, New York. 1965, pp. 38–61.

Chargaff, E. and R. Lipshitz. *J. Am. Chem. Soc.* 75: 3658 (1953).

Lehman, I. R., M. C. Bessman, E. Sims and A. Kornberg. *J. Biol. Chem.* 223: 163 (1958).

Meselson, M. and F. Stahl. *Proc. Natl. Acad. Sci. U.S.* 44: 461 (1958).

Schwimmer, S. *Phytochemistry* 5: 791 (1966).

Watson, J. D. and F. H. C. Crick. *Nature* 171: 737 (1953).

Watson, J. D. and F. H. C. Crick. *Cold Spring Harb. Symp. Quant. Biol.* 18: 123 (1953).

EXPERIMENT 14 ISOLATION OF DNA

A. OBJECTIVE

In eucaryotic organisms, DNA is primarily located in the nucleus. The DNA of the chromosomes is associated with a great deal of many types of proteins. Histones and other basic and nonbasic proteins bind to DNA to form a nucleoprotein structure. Isolation techniques are designed to separate the DNA from protein both chemically and physically. This experiment is planned to allow the student to become acquainted with this procedure.

B. EQUIPMENT AND SUPPLIES

Plant tissue (lily bulbs or leaves)
Dupanol-EDTA-SSC (5% Dupanol, 0.001 *M* EDTA, 0.15 NaCl, 0.015 *M* sodium citrate, pH 7.4)
Cold ethanol
Chloroform–isoamyl alcohol
SSC (0.15 *M* NaCl and 0.015 *M* sodium citrate, pH 7.6)
Homogenizer
Centrifuge
Diphenylamine reagent (see Experiment 11)
Spectrophotometer
Water bath

C. EXPERIMENTAL PROCEDURE

1. Homogenization and DNA Extraction

Chop 100 g of fresh plant tissue (choose a plant material of your interest; usually plants of the lily family are best) into small pieces (about 1 mm^3) and then add 100 ml of Dupanol-EDTA-SSC (5% Dupanol, 0.001 *M* EDTA, 0.15 *M* NaCl, 0.015 *M* Na citrate, pH 7.4), and incubate at 60 C for 15 min. Then quickly cool to room temperature and homogenize in a Vir-Tis (or similar) homogenizer in this manner: 30 sec, medium speed; 1 min, low speed; 1 min, high speed; and 1 min, low speed. Let the homogenate stand for 1 hr at room temperature and then centrifuge at 12,000 × *g* for 15 min. Collect and save the supernatant. Add 50 ml of Dupanol-EDTA-SSC to the precipitate, mix, and then incubate the mixture at room temperature for 30 min. Centrifuge the mixture and collect the supernatant. Pool the two supernatants and add NaCl (solid) to make the final concentration equal to 1.15 *M*.

2. Deproteinization

Add an equal volume of chloroform–isoamyl alcohol (20:1) to the combined Dupanol-EDTA-SSC supernatant. Shake this mixture well from time to time for about 5 min. Centrifuge to separate the two immiscible liquids and then remove the aqueous layer (top) with a large hypodermic syringe and needle. Leave behind the layer of protein at the interphase. Repeat the deproteinization step until no protein precipitate can be seen at the interphase.

3. Alcohol Precipitation

Add 2 volumes of cold ethanol to the deproteinized aqueous solution to precipitate DNA. After centrifugation at 10,000 × *g* for 15 min decant and discard the supernatant (drain tubes well). Dissolve the pellet in 10 ml of SSC (0.15 *M* NaCl, 0.015 *M* Na citrate, pH 7.6).

4. DNA Spooling

Remove the DNA by spooling out the fibrous material on a rough stirring rod. This can be achieved by layering cold ethanol over the 10 ml of DNA solution sample (in a test tube). While adding the ethanol (no more than 2 volumes), gently mix (swirl action) the solution with a glass stirring rod. Spool the fibrous material around the stirring rod as the precipitate forms. DNA should spool out as the Na^+ salt; see (1). In case the DNA cannot be spooled out, add the equivalent of 2 volumes of cold ethanol and collect the precipitate by centrifugation.

5. Analysis

Dissolve a sample of the purified DNA in SSC and determine the U.V. absorbancy spectrum with a spectrophotometer. Usually 1 OD unit equals 50 μg of DNA. Also measure the amount of DNA by the diphenylamine test for deoxyribose (see Experiment 11). From both the U.V. absorbancy data and the diphenylamine test calculate the DNA yield.

6. Fractionation of DNA on an MAK Column (Optional)

Usually DNA isolated from plants is contaminated with a large amount of RNA. Removal of the RNA from DNA may be achieved by fractionation of a sample on an MAK column (see Experiment 19). Add approximately 45 OD units of DNA to the MAK column and elute with an NaCl gradient from 0.3 to 0.8 *M*. Calculate the DNA yield by summing the OD in the DNA peak × volume. *Store the fibrous DNA* (purified and nonpurified) *in the freezer for later experiments.*

D. TREATMENT OF DATA

Prepare a short report of your data and comment on the procedure in terms of how it could be improved.

REFERENCE

1. Ergle, D. R. and R. H. Katterman. *Plant Physiol.* 36: 811 (1961).

General References

Bendick, A. J. and E. T. Bolton. *Plant Physiol.* 42: 959 (1967).
Marmur, J. *J. Mol. Biol.* 3: 208 (1961).

EXPERIMENT

15 A NEW METHOD FOR DNA PURIFICATION

A. OBJECTIVE

It is frequently desirable to obtain DNA of high purity, for a number of obvious reasons. The procedure presented in Experiment 14 often does not give DNA of high quality. However, the DNA obtained from that procedure is adequate for the determination of base composition or melting profiles. To obtain DNA for DNA-RNA hybridization, as a source of template for either DNA polymerase or RNA polymerase, etc., another method of preparing the DNA is needed. Recently a group (Britten, Pavich, and Smith) from the Carnegie Institute devised a new method for DNA purification. The objective of this experiment is to purify high-quality DNA by their method.

B. EQUIPMENT AND SUPPLIES

Plant tissue (corn shoots)
Homogenizer
Phosphate buffers (0.14 *M*, pH 7.0; 0.05 *M*, pH 7.0)
8 *M* Urea, 0.24 *M* sodium phosphate, pH 7.0, 0.01 *M* EDTA, 1% Dupanol
Hydroxylapatite
5% Octanol
Glass column (2 × 50 cm)
Centrifuge
Spectrophotometer
Peristaltic pump

C. EXPERIMENTAL PROCEDURE

1. Preparation of Cell Homogenate

Homogenize 50 g of etiolated corn shoots (5 days after planting) with 100 ml of 0.14 *M* sodium phosphate buffer, pH 7.0, in a Vir-Tis (or similar) homogenizer for 30 sec at medium speed followed by 45 sec at high speed. Squeeze the homogenate through four layers of cheesecloth. Then centrifuge the filtrate at 7000 × *g* for 30 min to pellet the chromosomal material. Decant and discard the supernatant. Suspend the pellet in about 20 ml of a solution containing 8 *M* urea, 0.24 *M* sodium phosphate, 1% Dupanol, and 0.01 *M* EDTA, pH 7.0. Next homogenize this suspension in a homogenizer at low speed, 5–10,000 rpm, for about 1 min. Save this solution for purification on hydroxylapatite. To improve yield, it may be advisable to treat the supernatant with 1 *M* $NaClO_4$ at room temperature for a few minutes. Then extract this solution in a separatory funnel with an equal volume of chloroform containing 5% octanol. Save the aqueous phase for the hydroxylapatite column.

2. Preparation of the Hydroxylapatite Column

To make the hydroxylapatite column first prepare:

(a) 0.05 *M* Sodium phosphate buffer, pH 7.0.

(b) A suspension of 10 g hydroxylapatite (Bio-Gel HTP from Bio-Rad Laboratories), and 1 g of cellulose powder (CF 11 from Whatman) in 60 ml of phosphate buffer (a).

(c) A suspension of two portions of 0.5 g of cellulose powder in 10 ml of phosphate buffer (a) each.

Pour about 30 ml of phosphate buffer into a column (2 × 50 cm) fitted with a sintered-glass bottom. Prevent the liquid from flowing through the column by using a stopcock or pinchclamp. Then add one of the 0.5-g cellulose suspensions (c) to the column. After the cellulose settles, add the suspension of hydroxylapatite and cellulose to the column (stir well before adding). Open the stopcock and allow the liquid to flow rapidly from the column. Finally, pour the second portion of cellulose into the column and allow the material to settle. Before using the packed column, equilibrate it by washing with 200–300 ml of 0.05 *M* phosphate buffer. This can be done easily by using an ordinary peristaltic pump with a flow rate of about 10 ml/hr.

3. Fractionation of DNA on Hydroxylapatite (HAP)

Pour the crude supernatant or the $NaClO_4$-treated aqueous phase onto the HAP column to adsorb the DNA to the HAP. Wash the extract onto the column with 0.24 *M* phosphate buffer, pH 7.0, containing 8 *M* urea and 0.4% Dupanol (use several volumes until the OD of the eluate is almost to zero). Then wash the column with several volumes of 0.014 *M* phosphate buffer, pH 7.0, to remove the urea. A refractometer, if available, is useful for checking salt concentrations and the removal of urea.

The DNA is eluted from the HAP column with 0.4 *M* phosphate buffer. All these procedures should be performed at room temperature. Collect 3- to 5-ml fractions from the HAP column and read the OD at 260 mμ. Save the peak fractions. The phosphate buffer may be removed by dialyzing the DNA solution against H_2O. If dialysis dose not remove the phosphate or the solution becomes too dilute, the DNA may be precipitated by adding 2 volumes of cold ethanol. The precipitate can be collected by centrifugation and then dissolved in a small amount of buffer or H_2O.

4. Analysis of the DNA

Determine the U.V. absorption spectrum of the purified DNA. Also determine the DNA content by the diphenylamine method (see Experiment 11). If the equipment is available, a melting profile could be determined on the DNA. Alternatively, an easy method of obtaining an estimate of the double-strandedness of the DNA is: Take a sample of DNA and add NaCl to give a final Na^+ concentration of 0.2 *M* (including the Na^+ from the phosphate buffer). After determining the initial OD at 260 mμ, heat the DNA solution for 5 min in boiling water. Place the tube containing the DNA in ice until the solution cools to near room temperature. Then measure the absorbancy at 260 mμ and determine the percent increase in absorbancy due to heating.

D. TREATMENT OF DATA

Present your data in the form of a research report to some research-supporting agency. Comment on the usefulness of the method for plant tissues and the work that should be done to improve the method.

GENERAL REFERENCE

Carnegie Institution Yearbook. Washington, D.C. 1970, pp. 400–402.

EXPERIMENT

16 DENSITY GRADIENT SEPARATION OF DNA

A. OBJECTIVE

The density of DNA is a function of its base composition. The higher the GC content (or the lower the AT content), the greater is the density.

Therefore DNAs of different base ratios can be separated by centrifuging the DNA mixture in CsCl at high speed. A gradient of CsCl can be established manually or by centrifugation. In either case, when a DNA sample is placed on top of a CsCl gradient and centrifuged for several hours, the DNA will band in the gradient where the densities of the CsCl and DNA are the same.

The objective of this experiment is to learn density gradient techniques and to separate two samples of DNA having different base compositions.

B. SUPPLIES AND MATERIALS

DNA (*E. coli* and salmon sperm)
CsCl (65% w/v)
Ultracentrifuge
Gradient maker
Stirring apparatus
Gradient collector
Paraffin oil

C. EXPERIMENTAL PROCEDURE

1. Preparation of CsCl Gradient

Using three different gradients, separate two DNA samples (*E. coli* and salmon sperm) by gradient centrifugation. First, prepare a solution of 65% CsCl by dissolving 10 g of CsCl in 5.4 ml of 0.01 *M* Tris buffer, pH 9.0 (this stock CsCl solution will be used to make other concentrations by dilution of the stock solution with the Tris buffer). Using a special gradient maker, prepare the following CsCl gradients: 45–65%, 48–63%, and 50–60%. Set up the gradient maker and other apparatus as illustrated in Figure 41.

When the apparatus is set up, add 2 ml of the more concentrated CsCl (for example, 65%) to chamber A and 2 ml of the less concentrated (for example, 45%) solution in chamber B. Start the stirrer and open the valve between the two chambers. Then open the valve to allow the solution to flow from chamber A through the hypodermic needle. Regulate the flow so that the gradient will be finished in about 10 min. Make sure that the hypodermic needle is touching the top inside wall of the tube so that the CsCl will flow smoothly to the bottom without forming drops.

Repeat the process to form the other two gradients without disturbing the finished gradient. Then add 0.2 ml each of *E. coli* and salmon sperm DNA (5 mg/ml dissolved in 0.01 *M* Tris buffer, pH 9.0) to the top of the gradient. Next add 0.2 ml of paraffin oil and carefully place the tubes in the SW 39.1 Spinco rotor or equivalent ultracentrifuge rotor.

FIGURE 41. An apparatus to make linear gradients for ultracentrifugation.

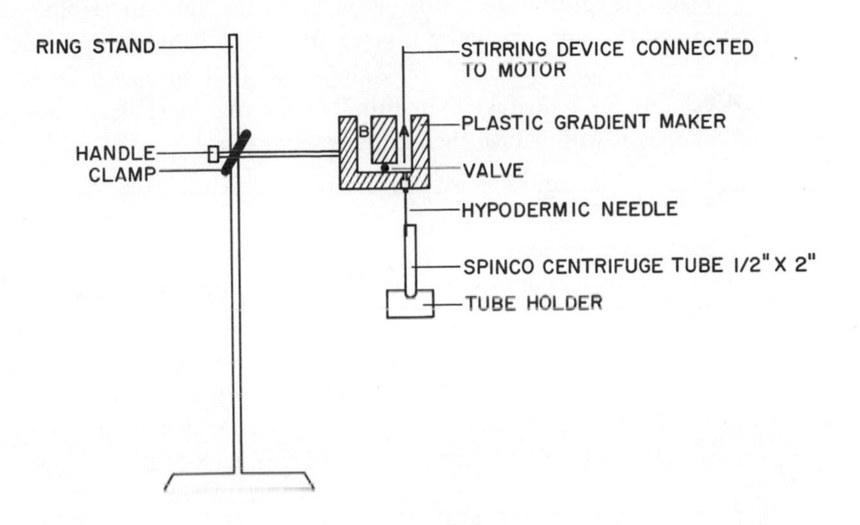

FIGURE 42. A simple gradient collector (a method of collecting a solution from a centrifuge tube).

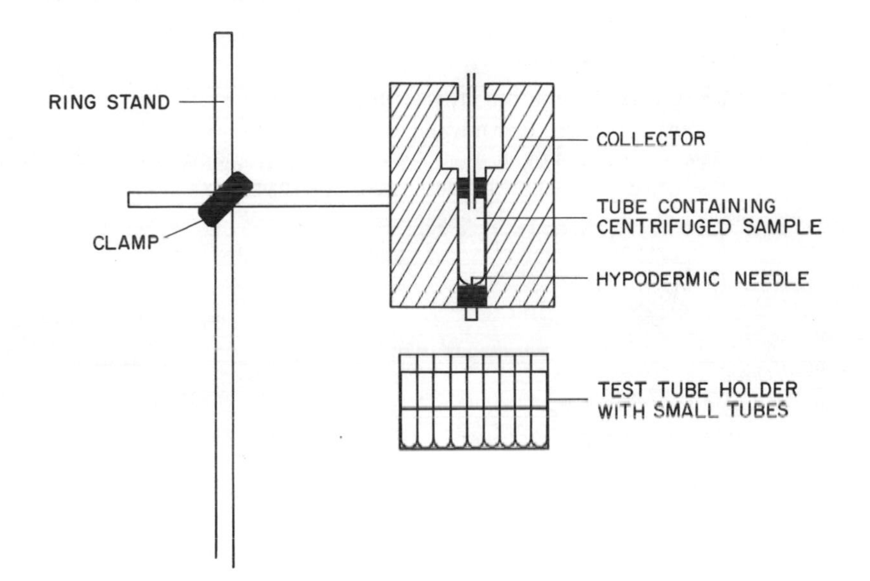

2. Centrifugation

Place the rotor in the centrifuge, and set the timer to "Hold" and the rpm dial to zero. Gradually (over a period of 20 min) bring the speed to 32,500 rpm. Centrifuge the samples at that speed for 20 hr. Then decelerate the centrifuge by manually reducing the rpm. Do not use the brake, as it will disturb the gradient separations.

3. Collection of Samples

Samples are taken from each tube by puncturing the bottom of the tube with a hypodermic needle fitted into another plastic apparatus, as illustrated in Figure 42.

Manually collect three drops per fraction into test tubes placed below the collector. Add 1 ml of water to each tube and determine the U.V. absorbancy at 260 mμ, using cuvettes which hold 1 ml or less. Plot the OD against the fraction number for each of the three gradients.

D. TREATMENT OF RESULTS

Present your results in a short communication. Assuming that the gradients produced were linear, estimate the densities of the two DNAs.

GENERAL REFERENCES

Cherry, J. H. *Science* 146: 1066 (1964).

Marmur, J., R. R. Rownd and C. L. Schildkraut. In *Progress in Nucleic Acid Research*. Vol. 1. Eds. J. N. Davidson and W. E. Cohn. Academic Press, New York and London. 1963, pp. 231–300.

Sober, H. A. and R. A. Harte, eds. *Handbook of Biochemistry: Selected Data for Molecular Biology*. Chemical Rubber Co., Cleveland, Ohio. 1968, pp. G-1 to H-65.

EXPERIMENT 17 EXTRACTION AND ASSAY OF MAMMALIAN DNA POLYMERASE

A. OBJECTIVES

DNA replication is a fundamental prerequisite to cell division. In order that the genetic information be exactly duplicated for each daughter cell, the enzyme, DNA polymerase, copies each DNA strand to yield a pair of identical replicates. The objective of this experiment is for the student to learn techniques dealing with the incorporation of radioactive deoxynucleoside triphosphates into DNA.

B. SUPPLIES AND MATERIALS

Beef liver (or rat liver)
Homogenizer
Centrifuge
Buffers (A: 0.02 *M* Tris-HCl, pH 7.5, 0.001 *M* EDTA, and 0.005 *M* 2 mercaptoethanol; B: 0.02 *M* Tris-HCl, pH 7.5, 0.15 *M* KCl, 0.01 *M* EDTA, and 0.005 *M* 2-mercaptoethanol)
Fine glass beads
2 *N* Acetic acid
0.2 *M* Perchloric acid
Deoxynucleoside triphosphates
Heated calf thymus DNA
^{3}H-dATP
Mortar and pestle
Whatman glass filter discs (GF/A)
Alumina A-306
Microscope

C. EXPERIMENTAL PROCEDURE

1. Tissue Homogenization

Chill about 5 g of fresh beef liver or rat liver in ice water. Mince the tissue and homogenize it in 50–75 ml of buffer A (0.02 *M* Tris-HCl, pH 7.5, 0.001 *M* EDTA, and 0.005 *M* 2-mercaptoethanol) with a Potter-type homogenizer. All operations are to be performed at 0–4 C. Cell disruption should be over 75% complete and should be checked microscopically. Solid KCl is then stirred into the homogenate to give a final concentration of 0.15 *M*. The preparation is then centrifuged at 20,000 × *g* for 60 min in a refrigerated centrifuge to give a supernatant (S-1) and precipitate (P-1). With certain cells, polymerase activity may be retained in the P-1 fraction. To release the bulk of this activity suspend P-1 in 10–15 ml of buffer B (0.15 *M* KCl, 0.02 *M* Tris-HCl, pH 7.5, 0.01 *M* EDTA, 0.005 *M* 2-mercaptoethanol). Homogenize the suspension for 20 sec, then vigorously shake (Check under a microscope to ensure that cell nuclei are not broken) Repeat this step if cells are not broken. Centrifuge the preparation for 60 min at 20,000 × *g* to obtain P-2 and S-2. S-2 may be combined with S-1, but for some purposes it may be kept separate, since it may have a higher specific activity of polymerase. The approximate protein concentrations of S-1 and S-2 should be 3 and 4 mg protein/ml, respectively. The Warburg and Christian method (Experiment 7) may be used to determine protein concentration of the various fractions.

DNA polymerase from other cells (for example, from bean leaves, spinach leaves, *Chlorella*) may not be readily released. This is partly due to the toughness of the tissue or cell wall, and partly to the strong binding of the enzyme to intracellular structures. Successful disruption techniques in these cases include blender-type homogenization in the presence of fine glass beads; grinding in mortar with glass beads or acid-washed sea sand; and sonication.

2. *Enzyme Purification*

A rapid and simple five to tenfold purification of polymerase is effected by adjusting the pH of the S fraction to 4.8–5.0 by the addition of 2 *N* acetic acid while the solution is gently stirred. After 5 min, centrifuge at 10,000 × *g* for 15 min to give "pH 5-Sup" and "pH 5-Ppt." The latter contains 90% of the polymerase and only 10% of the DNase of the fraction. The pellet is dissolved without delay in buffer A (20% of volume of buffer taken for initial cell disruption).

For further purification, see Shepherd and Keir (1), Yoneda and Bollum (2), and Gold and Helleiner (3).

3. *Assay*

The assay is usually done with a 0.25-ml volume in a 3-ml stoppered tube. The assay mixture contains the following in 0.25 ml: 5 μmoles of Tris-HCl buffer, pH 7.5; 15 μmoles of KCl; 0.1 μmole of EDTA; 50–100 μg of heated DNA (90 C for 10 min); 5 μmoles each of dATP (^{3}H-labeled), dCTP, dGTP, and dTTP; 1–2 μmoles of $MgCl_2$ or $MgSO_4$; ca. 1 μmole of 2-mercaptoethanol; 100–150 μg of protein containing DNA polymerase. (A suitable amount of ^{3}H-dATP to add to the reaction is 10^6 d.p.m. per assay mixture.) The standard incubation is run at 37 C for 60 min. Thereafter the reaction mixture is cooled to 0 C (may be stored at −20 C) and treated in the following way.

To determine the amount of incorporation of radioactive precursor, add 1.5 ml of cold water, 0.2 ml of heated DNA (2.0 mg/ml), and 4 μmoles of unlabeled ATP to each tube. Six percent perchloric acid (0.5 ml) is added with mixing, and after a few minutes the mixture is shaken to induce flocculation. The suspension is filtered by suction through glass fiber discs (GF/A Whatman, 2.1 cm). Use Millipore glass holders for this procedure. Rinse the tube with 3 ml of cold 0.2 *N* perchloric acid and pour through the filter apparatus. Then wash the disc successively with four 10-ml portions of cold 0.2 *M* perchloric acid and with 100% ethanol. The dry filter is then placed in a vial containing scintillation fluid and the radioactivity is counted in the liquid scintillation spectrophotometer.

Results are conveniently expressed as mμmoles radioactive triphosphate incorporated/mg protein.

D. TREATMENT OF RESULTS

Calculate the specific activity of DNA polymerase activity in terms of mμmoles and, on the basis of 100% enzyme yield, determine the amount of DNA that 5 g of liver could produce per hr. Present the results in the form of a short report.

REFERENCES

1. Shepherd, J. B. and H. M. Keir. *Biochem. J.* 99: 443 (1966).
2. Yoneda, M. and F. J. Bollum, *J. Biol. Chem.* 240: 3385 (1965).
3. Gold, M. and C. W. Helleiner. *Biochim. Biophys. Acta.* 80: 193 (1964).

General Reference

Keir, H. M. *Prog. Nucl. Acid Res. Mol. Biol.* 4: 81 (1965).

EXPERIMENT 18 DNA SYNTHESIS BY SOYBEAN HYPOCOTYL CHROMATIN

A. OBJECTIVE

Associated with its DNA, chromatin contains various histones and nonhistone proteins. DNA polymerase is one of the enzymatic proteins associated with chromatin isolated from soybean hypocotyls. The objective of this experiment is to learn techniques for the isolation of chromatin and the assay of DNA polymerase.

B. SUPPLIES AND MATERIALS

Soybean seedlings (4 days old)
High-speed homogenizer
Dounce tissue grinder
Ultracentrifuge
Deoxynucleoside triphosphates
^{3}H-TTP (12 mCi/μmole)
0.01 *M* Hepes buffer, pH 7.5
Glass fiber filters (Whatman GF/A)
Miracloth

C. SOLUTIONS

All of these materials are at pH 8.0 (with HCl)

Grinding Medium	*Wash Medium 1*	*Sucrose*
0.25 *M* Sucrose	0.25 *M* Sucrose	2 *M* Sucrose
0.05 *M* Tris	0.01 *M* Tris	0.01 *M* Tris
0.001 *M* $MgCl_2$	0.01 *M* 2-Mercapto-ethanol	0.01 *M* 2-Mercapto-ethanol
0.015 *M* 2-Mercapto-ethanol		

D. EXPERIMENTAL PROCEDURE

1. Chromatin Isolation

Harvest 150 g of hypocotyl tissue from 4-day-old soybean seedlings (grown in the dark at 29 C). Homogenize the tissue in a Vir-Tis (or similar) homogenizer for 30 sec at medium speed followed by 45 sec at high speed, using 1.5 ml of grinding medium per gram of tissue. All operations are to be carried out at 0–4 C. Squeeze the homogenate through four layers of cheesecloth and filter it through Miracloth (Calbiochem). Then centrifuge the filtrate at 5000 × *g* for 30 min. Remove the supernatant with a syringe and discard it. Add approximately 10–15 ml of wash medium 1 to the pellet and gently suspend the gelatinous portion of the pellet with a round-tipped spatula, being careful not to remove any starch material. Decant the gelatinous material (chromatin) into a 40-ml Dounce tissue grinder equipped with a "B" pestle (Kontes Glass Company, Vineland, N.J.) and gently agitate until a homogeneous mixture is obtained. This suspension is centrifuged at 13,000 × *g* for 15 min and the gelatinous portion of the pellet is again resuspended in wash medium and recentrifuged. Repeat this step until no starch pellet is obtained. After the final wash, chromatin is resuspended in 5–15 ml of wash medium and layered over 20 ml of 2 *M* sucrose solution in centrifuge tubes for the Spinco SW 25.1 rotor (other centrifuges may be used). Gently stir the upper one-third of the 2 *M* sucrose solution with a spatula. Chromatin is pelleted by centrifugation at 20,000 rpm for 3 hr. Discard the supernatant, wipe the tube walls dry with absorbent paper, and then suspend the pellet in 1–2 ml of 0.01 *M* Hepes buffer, pH 7.5.

2. DNA Polymerase Assay

Labeled deoxyribonucleoside triphosphate is incorporated into trichloroacetic acid–insoluble material to measure DNA polymerase activity. The standard reaction mixture contains, in micromoles, the following: Hepes, pH 7.5, 20; $MgCl_2$, 1; dithiothreitol, 1; dATP, dCTP, dGTP, 0.05 each; TTP, 0.002; and 5 μCi of ^{3}H-TTP, and chromatin equivalent to 4–8 μg of DNA, in a final volume of 0.2 ml. Incubate the reaction mixture for 30 min at 37 C, and stop the reaction by adding 3 ml of 10% trichloroacetic acid containing 10^{-3} *M* inorganic pyrophosphate. Using a Millipore filter apparatus, transfer the precipitated material onto glass fiber filter discs (Whatman GF/A) and wash five times with 5-ml portions of cold 5% trichloroacetic acid. Then dry the filters under infrared lamps, place them in the scintillation fluid, and count the radioactivity in a liquid scintillation spectrometer. Measure the chromatin DNA content by the diphenylamine procedure of Burton (see Experiment 11) after initial hydrolysis in 0.5 *N* perchloric acid for 15 min at 70 C.

E. TREATMENT OF RESULTS

Calculate the incorporation of ^{3}H-TTP/100 μg DNA/hr. Present the results as a short report.

GENERAL REFERENCES

Holm, R. E., T. J. O'Brien, J. L. Key and J. H. Cherry. *Plant Physiol.* 45: 41 (1970).

Huang, R. C. C. and J. Bonner. *Proc. Natl. Acad. Sci. U.S.* 48: 1216 (1962).

Jarvis, B. C., B. Frankland and J. H. Cherry. *Plant Physiol.* 43: 1734 (1968).

Leffler, H. R. Ph.D. thesis, Purdue University, August 1970.

Leffler, H. R., T. J. O'Brien, D. V. Glover and J. H. Cherry. *Plant Physiol.* 48: 43 (1971).

Schwimmer, S. *Phytochemistry* 5: 791 (1966).

Stout, E. R. and M. Q. Arens. *Biochim. Biophys. Acta* 213: 90 (1970).

PART

8

RIBONUCLEIC ACID

INTRODUCTION

The genetic information of DNA is transferred to another class of molecules, which then serve as the templates of protein synthesis. These intermediate templates are molecules of ribonucleic acid (RNA) and are chemically very similar to DNA. The relation of DNA to RNA to protein is usually called the central dogma and is illustrated in the following manner:

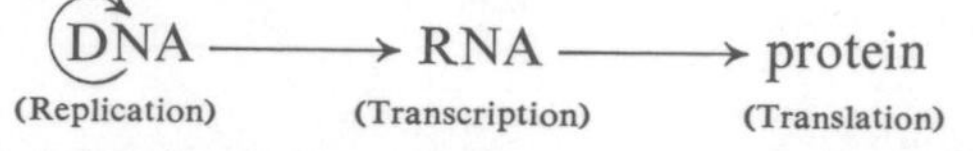

The function of nucleic acids as templates for protein synthesis is now fairly well understood. The complementarity of the nucleotide sequence of messenger RNA and DNA can be tested by various DNA-RNA hybridization techniques. RNA sequences complementary to DNA form stable DNA-RNA hybrids which can be easily detected. Spiegelman and Hayashi (1), using the hybridization technique, showed that the RNA synthesized in bacterial cells after bacteriophage infection is complementary to the phage DNA. These and subsequent experiments have given credence to the idea that messenger RNA molecules are transcribed from the DNA template. However, it is to be noted that in a few unusual cases certain viral RNA directs DNA polymerization. Messenger RNA transcribed from DNA is then translated by the "protein synthesis system" into new polypeptides.

In addition to the so-called messenger RNA, several other types of RNA are transcribed from the DNA template. These types include soluble or transfer (4S), 5S ribosomal, and the two major ribosomal (18S and 25S for plants) RNAs. With the exception of 5S RNA, all the other types participate in protein synthesis, which is discussed in Part 9. Recently there has been evidence for new types of RNA. One, chromosomal RNA, is found in low concentrations exclusively in the nuclei. Data from Bonner's laboratory (2) show that this class of RNA molecule is composed of 50–60 nucleotides. A characteristic of this RNA is that it contains the unusual nucleotide, dihydrouridylic acid. The RNA is thought to participate in directing the interaction of histone with DNA.

Another class of RNA which is found in the nucleus may be termed "regulator RNA." In cells of higher organisms a large amount of RNA is rapidly synthesized in the nucleus and remains there. Britten and Davidson (3) have proposed that this type of RNA is the product of sensor genes and that it regulates the transcription of the structural (producer) genes. At the moment, little is known about this type of RNA. It is clear, however, that the regulation of gene transcription in higher organisms is quite different from that of bacterial cells.

RNA ISOLATION

The main objective in the isolation of RNA is to obtain a high yield of undegraded polymer. The techniques used most frequently originated from the work of Kirby (4). As RNA of the living cell is usually complexed with protein, the extraction process involves the removal of RNA from protein primarily by disruption of hydrogen bonds. Kirby has shown that this can be achieved with an aqueous phenol solution containing a detergent such as sodium lauryl sulfate (Dupanol). In many organisms, especially higher plants, high levels of RNase activity require that appropriate inhibitors be added and/or the enzyme be removed during RNA isolation. A negatively charged clay (bentonite) is often added to the extraction medium to inhibit RNase activity. In some cases other materials such as naphthalene-1,5-disulfonate, 8-hydroxyquinoline, and *m*-cresol are used to break hydrogen bonds and therefore inhibit RNase.

At pH's near neutrality, RNA is soluble in polar solvents. Therefore, when tissues are homogenized in buffers containing phenol, detergents, bentonite, or other materials and the homogenate is then centrifuged, the material is distributed in the tube as shown in Figure 43.

The aqueous phase is removed from the tube, and the RNA is precipitated with cold ethanol. The precipitate containing both RNA and DNA is usually dialyzed or processed by some column technique to remove contaminating materials. When a detergent is omitted from the extraction medium, most of the DNA and a large amount of the newly synthesized RNA are contained in the gel interphase. Both types of nucleic acids can be differentially extracted by suspending the gel interphase in a phenol-buffer solution containing detergent (5). In fact, several workers have used this technique to extract selectively DNA and the rapidly metabolized RNA.

Another technique of selectively isolating RNA deals with the extraction of soluble (s)RNA in 1 *M* NaCl or in 2–3 *M* potassium acetate for plants. Thus sRNA can be solubilized from ethanol precipitates of total nucleic acid fractions, whereas rRNA is insoluble.

FIGURE 43. Distribution of an aqueous phenol extract after centrifugation.

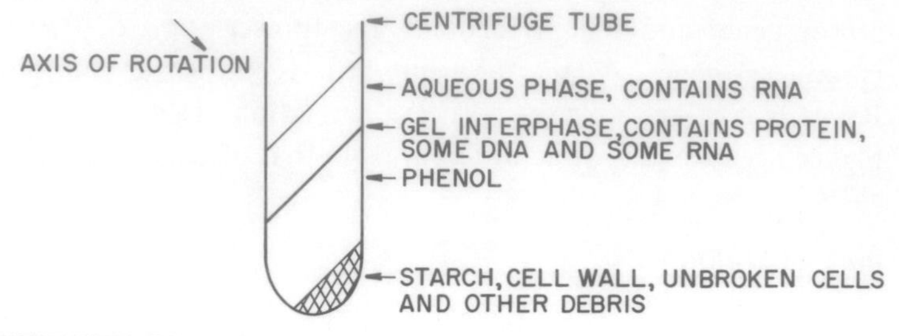

PROPERTIES OF RNA

RNA consists of four ribonucleosides held together by phosphate esters at the 3′ and 5′ positions of ribose. As previously discussed (Part 6), RNA is degraded to (2′)3′-nucleoside monophosphates by alkali. A number of nucleases attack RNA, depending on phosphate linkage and base sequence. Since RNA is a strong polyanion, most of the RNA in cells occurs in the form of some complex, usually with proteins. With the exception of tRNA, RNA usually occurs as single strands. Transfer RNA, on the other hand, shows a considerable interstrand base complementary structure. Consequently, tRNA has only a few bases which are not base-paired. A good review of the general properties of RNA is provided by Markham (6).

RNA SYNTHESIS

Four different enzyme systems are involved in the polymerization of ribonucleotides into RNA. The true "transcriptase" is RNA polymerase, which is the most important enzyme involved in RNA synthesis in bacteria and higher organisms. However, polynucleotide phosphorylase (7), an enzyme that catalyzes the synthesis of RNA from nucleoside diphosphates, is found in a wide range of organisms. Another enzyme, the viral replicase, is primarily involved with the synthesis of complementary viral RNA. Probably the system least understood in terms of physiological importance is the enzymes that catalyze the synthesis of polyribonucleotide homopolymers (8). Because of the great importance of RNA polymerase in the synthesis of most of the RNA in cells, only this system is discussed further.

Escherichia coli DNA-dependent RNA polymerase is composed of at least five different subunits, each playing a different role in *in vivo* transcription of DNA (9). Electrophoresis of purified RNA polymerase on polyacrylamide gels containing denaturing agents such as sodium dodecyl sulfate or urea gives a number of polypeptide chains. The possible functions and molecular weights of the subunits are listed in

TABLE 9

Subunit	*Molecular weight*	*Number/mole*	*Function*
β'	165,000 ± 10%	1	DNA binding
β	155,000 ± 10%	1	σ and DNA binding
σ	95,000 ± 5%	1	Initiation
α	39,000 ± 5%	2	?
ω	9,000 ± 10%	1	?

Table 9. Thus it appears that the *E. coli* polymerase has an exceedingly complex subunit structure, but at present it is not known why such a complex structure is required or what the function of each subunit is.

Chromatography of purified polymerase on phosphocellulose yields two peaks: a core polymerase which contains the machinery for RNA synthesis, but is deficient in its ability to initiate such synthesis efficiently and accurately on certain DNA templates; and a σ factor which has no synthetic activity, but, when added back to the core polymerase, forms the complete enzyme and restores the ability to initiate properly. The σ factor has been shown to act catalytically in stimulating initiation of RNA chains, and once initiation has occurred it is released from the core enzyme. In addition to its effect on initiation, the σ factor also affects the aggregation of the enzyme at low ionic strength and the dissociation into inactive intermediates on treatment with mild denaturants.

The finding that the σ factor is required for specific initiation and the discovery of several bacteriophage-coded σ-like factors needed for the initiation of certain viral-specific RNA syntheses have led to the suggestion that such initiation specificity might play a major role in the regulation of RNA synthesis. These factors would exert a general positive control, determining which genes could be transcribed by the core enzyme [see (10) for excellent reviews].

Reports of protein factors isolated from higher plants which enhance the activity of RNA polymerase have been made in at least two

FIGURE 44. A hypothetical model of auxin regulation of RNA transcription.

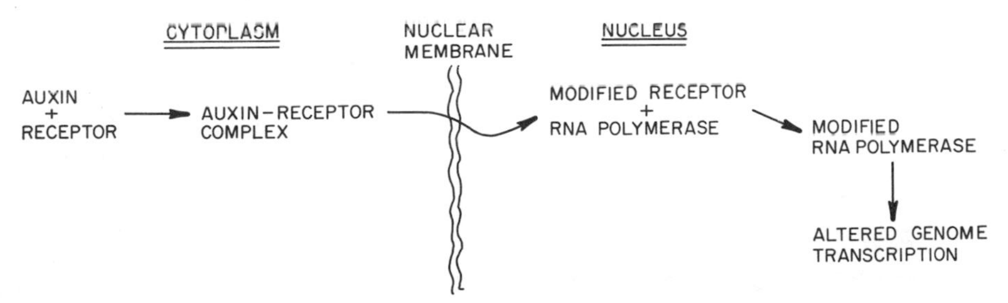

laboratories (11, 12). Several hormone-mediated responses could involve a receptor that in turn imposes certain regulations on RNA polymerase. Transcription of a battery of genes could, therefore, be regulated by a single receptor, as indicated in Figure 44, a model for the mode of action of auxin.

REFERENCES

1. Spiegelman, S. and M. Hayashi *Cold Spring Harb. Symp. Quant. Biol.* 28: 161 (1963).
2. Huang, R. C. and J. Bonner. *Proc. Natl. Acad. Sci. U.S.* 54: 960 (1965).
3. Britten, R. J. and E. H. Davidson. *Science* 165: 349 (1969).
4. Kirby, K. S. *Biochem. J.* 66: 495 (1957).
5. Cherry, J. H. and H. Chroboczek. *Phytochemistry* 5: 411 (1966).
6. Markham, R. In *Modern Methods of Plant Analysis*. Vol. 4. Eds. S. P. Colowick and N. O. Kaplan. Academic Press, New York. 1955, p. 246.
7. Grunberg-Manago, M. In *Progress in Nucleic Acid Research.* Vol. 1. Eds. J. N. Davidson and W. E. Cohn. Academic Press, New York. 1963, pp. 93–133.
8. Duda, C. T. and J. H. Cherry. *J. Biol. Chem.* 246: 2487 (1971).
9. Silvestri, L., ed. *RNA Polymerase and Transcription.* North-Holland Publishing Co., New York 1970.
10. *Cold Spring Harb. Symp. Quant. Biol.* 35 (1970).
11. Hardin, J. W., T. J. O'Brien and J. H. Cherry. *Biochim. Biophys. Acta* 244: 667 (1970).
12. Matthysse, A. G. and C. Phillips. *Proc. Natl. Acad. Sci. U.S.* 63: 899 (1969).

General References

Ewing, E. E. and J. H. Cherry. *Phytochemistry* 6: 1319 (1967).
Zubay, G. *J. Mol. Biol.* 4: 347 (1962).

EXPERIMENT 19 EXTRACTION AND CHROMATOGRAPHY OF TOTAL RNA

A. OBJECTIVE

The techniques for extraction of RNA from plants are designed to obtain as much of the RNA as possible from other contaminating macromolecules without degrading the RNA polymers by physical action or through the activity of RNase. The objective of this experiment is to provide the student with experience in the extraction and chromatographic separation of RNA.

B. EQUIPMENT AND SUPPLIES

Plant tissue (soybean or pea cotyledons)
Homogenizing buffer (0.01 *M* Tris, pH 7.6, 0.06 *M* KCl, 0.01 *M* $MgCl_2$)
11% Dupanol (sodium lauryl sulfate)
Bentonite (40 mg/ml)
Cold phenol (washed with homogenizing buffer)
Potassium acetate
Homogenizer
95% Ethanol
Refrigerated centrifuge
Dialyzing tubing
Glass column (2 × 40 cm)
0.05 *M* Sodium phosphate, pH 6.7
Fraction collector
Bovine serum albumin, fraction V (to make methylated albumin)
Spectrophotometer
Gradient maker
Kieselguhr
Cellulose powder

C. EXPERIMENTAL PROCEDURE

1. Extraction of RNA

Homogenize 15 g of tissue (soybean or pea cotyledons of 4-day-old seedlings) in a solution containing 30 ml of 0.01 *M* Tris, pH 7.6, 0.06 *M* KCl, 0.01 *M* $MgCl_2$; 1 ml of bentonite (40 mg/ml); 4.6 ml of 11% Dupanol and 50 ml of phenol* with a high-speed homogenizer (Vir-Tis). Homogenize for 1 min at full speed. Centrifuge the homogenate at 20,000 × *g* for 10 min. Carefully remove the aqueous layer, which contains the RNA, with a large syringe and needle. Notice that the precipitated protein is at the interphase of the phenol and aqueous phase. Add 1 ml of bentonite and 1 volume of phenol (washed twice with homogenization buffer) to the separated aqueous phase, and shake occasionally for about 5 min in the cold. Centrifuge this mixture at 20,000 × *g* for 10 min and again remove the aqueous layer. Reextract the aqueous phase with a half-volume of phenol at 0 C for 5 min. Centrifuge at 20,000 × *g* for 10 min and remove the aqueous layer. At this stage, very little protein should be noted at the aqueous interphase. Add solid potassium acetate to the aqueous solution to a final concentration of 0.2 *M* (approximately 20 mg/ml of solution), and then add 2 volumes of cold 95% ethanol. After a few minutes at ice-cold temperatures collect the RNA precipitate by centrifugation at 30,000 × *g* for 20 min. Decant the supernatant and dissolve the RNA pellet in 5–10 ml of 0.05 *M* sodium phosphate buffer, pH 6.7. Place the RNA solution in dialysis tubing and dialyze for at least 12 hr against 500 ml of 0.05 *M* sodium phosphate buffer, pH 6.7, in the cold with two

* Extreme caution must be taken when using phenol, as it burns the skin. If phenol accidentally gets on hands or face, immediately wash with water or ethanol.

changes of buffer at 4-hr intervals. After dialysis, carefully open the tubing and collect the dialyzed solution in a test tube. Dilute 0.1 ml of the RNA solution to 5 ml with 0.05 *M* sodium phosphate buffer, and determine the absorption spectrum from 220 to 300 mμ, taking readings at 5-mμ intervals.

2. Separation of Purified RNA on a Methylated Albumin Column

The methylated albumin kieselguhr (MAK) column used in this experiment is prepared according to the method of Mandell and Hershey (1) in the following manner:

(a) Methylation of Protein. Suspend 5 g of bovine serum albumin (fraction V from Sigma Chemical Co.) in 500 ml of absolute methanol. Add 4.2 ml of concentrated HCl, tightly seal the bottle, and incubate at 37 C for 5 days in the dark.

At the termination of the incubation period collect the precipitate by centrifugation and wash with 500 ml of methyl alcohol twice to remove HCl. The removal of HCl should be done as rapidly as possible to prevent hydrolysis of the methyl ester. After the methyl alcohol washes, the precipitate should be washed with anhydrous ether. Evaporate most of the ether in the air (*use a hood*) and store the powdered material under reduced pressure over KOH. A 1% solution of the methylated albumin is made in H_2O for the column. (The 1% MA should have a pH not lower than 3.5.)

(b) Preparation of MA and MAK Solutions. For the preparation of the MAK column, two forms of the methylated albumin are required, 1% MA and MA-coated kieselguhr (MAK). MAK is prepared as follows: Boil (to expel air) and then cool a suspension of 20 g of kieselguhr in 100 ml of 0.05 *M* sodium phosphate, pH 6.7, containing 0.1 *M* NaCl. Add 5 ml of 1% MA, stir, and add 20 ml more of the phosphate buffer—0.1 *M* NaCl. This suspension of kieselguhr coated with methylated albumin is designated MAK. *The stock solutions of MA and MAK will be prepared for the class.*

(c) Preparation of the Column. The MAK column used for analysis of nucleic acids is composed of four layers in a glass tube (2 × 40 cm) and is prepared as follows: In two beakers, boil and cool suspensions of kieselguhr as follows: (1) 8 g of kieselguhr in 40 ml of 0.1 *M* NaCl–0.05 *M* sodium phosphate, pH 6.7; (2) 6 g of kieselguhr in 40 ml of 0.1 *M* NaCl–0.05 *M* sodium phosphate, pH 6.7. To the first beaker add 2 ml of 1% MA and stir. To the second beaker add 10 ml of MAK and stir. The first layer of the column is made by suspending 1 g of standard grade cellulose powder in 20 ml of 0.1 *M* NaCl–0.05 *M* sodium phosphate, pH 6.7, and pouring it into a 2 × 40-cm glass column fitted with a porous glass disc on the bottom. Let the liquid drip from the column until the liquid level is nearly to the pad of powder. Before all of the solution drains from the column, form the second layer of the column by gently adding the contents of beaker 1 to the column with a pipet. Wash down the excess MAK with 0.1 *M* NaCl in phosphate buffer. As the solution drains from the column, the third layer of the column is formed by adding the contents from beaker 2 in a similar manner. The final layer of the column is made by adding the 1 g kieselguhr in 20 ml of 0.1 *M* NaCl and washing any excess kieselguhr down with the same solution. As a guide, a schematic diagram of the MAK column is provided in Figure 45.

(d) Running the Column. After the column is prepared, add about 35 OD_{260} units* of RNA (1.75 mg) in 40 ml of starting NaCl phosphate buffer (0.35 *M* NaCl–0.05 *M* sodium phosphate, pH 6.7) to the column. When the liquid level nearly reaches the kieselguhr, add 20 ml of starting buffer and let it flow through the column. When this liquid level nearly reaches the kieselguhr, add an additional 10 ml of starting buffer and set up the column to a gradient maker and a fraction collector as shown in Figure 46.

The RNA will be eluted from the column with a linear gradient of NaCl from 0.35 *M* to 1.1 *M* in 0.05 *M* sodium phosphate, pH 6.7. The gradient will be formed by adding 400 ml of 0.35 *M* NaCl–0.05 *M* sodium phosphate, pH 6.7, to the mixing chamber and 400 ml of 1.1 *M* NaCl–0.05 *M* sodium phosphate, at pH 6.7, to the reservoir. (The salt concentration may vary, depending on the degree of methylation of the protein. Therefore check with the instructor before starting the gradient.) When all components of the system are tightly connected and the magnetic stirrer is operating, apply 1.5 lb (psi) air pressure to the gradient maker, open the valve between the two chambers of the gradient maker, and collect the eluate from the column in 5-ml fractions. Determine the U.V. absorbancy of each fraction at 260 mμ. When the salt gradient is complete, and *before the column goes dry*, add 100 ml of 1.5 *N* NH_4OH to the column to remove the "tenaciously bound nucleic acid." Again collect 5-ml fractions but *do not* determine the OD at 260 mμ, as the fraction will contain the methylated protein removed from the column. A general elution profile of radioactive RNA is depicted in Figure 47.

* One OD_{260} unit is equal to that amount of material dissolved in 1 ml, necessary to absorb OD at 260 mμ in a 1-cm light path. Hence 3 ml of a solution in a cuvette which absorbs 1 OD at 260 mμ is equal to 3 OD_{260} units.

FIGURE 45. A schematic diagram of an MAK column.

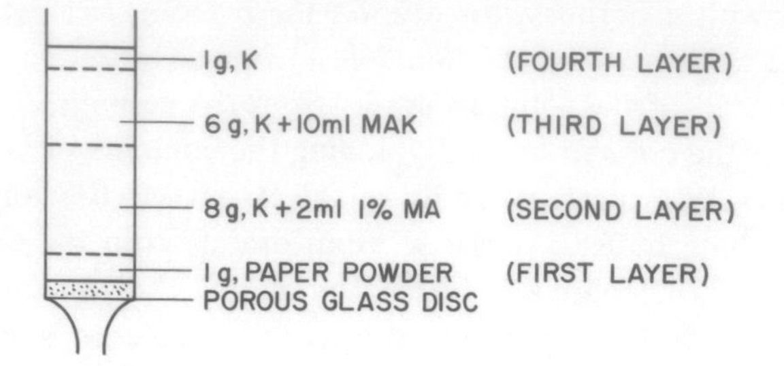

FIGURE 46. The gradient system and column required for MAK chromatography.

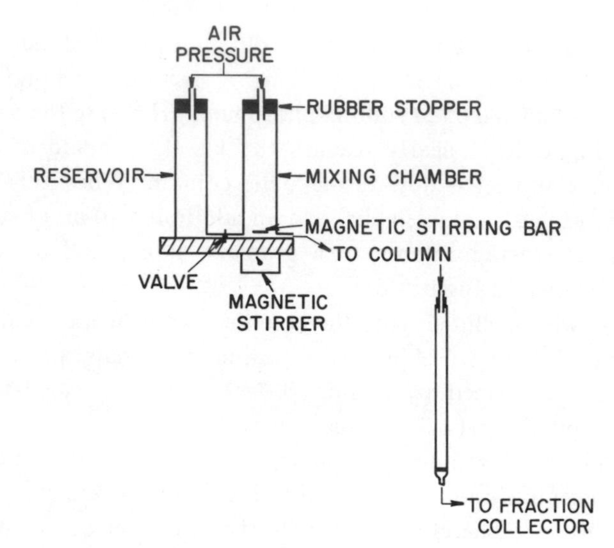

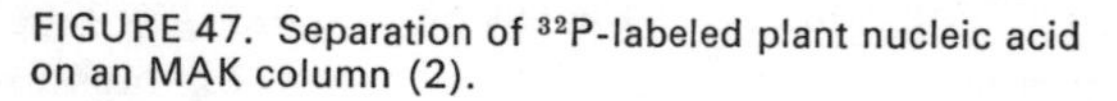

FIGURE 47. Separation of ^{32}P-labeled plant nucleic acid on an MAK column (2).

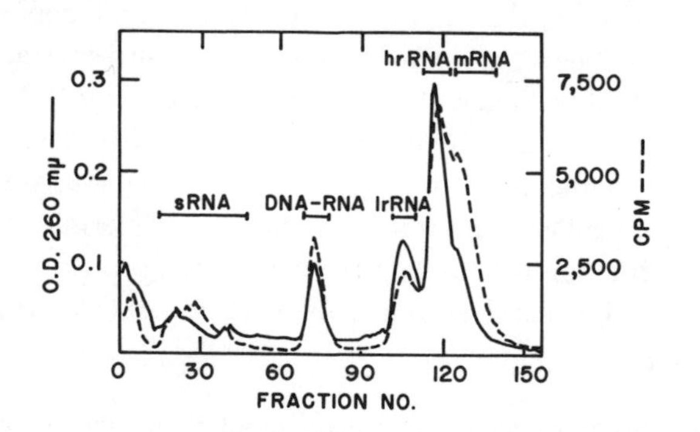

D. TREATMENT OF DATA

Summarize your data in figures, graphs, or tables and write a short report.

REFERENCES

1. Mandell, J. D. and A. D. Hershey. *Anal. Biochem.* 1: 66 (1960).
2. Ewing, E. E. and J. H. Cherry. *Phytochemistry* 6: 1319 (1967).

General References

Cherry, J. H., H. Chroboczek, W. J. G. Carpenter and A. Richmond. *Plant Physiol.* 40: 582 (1965).
Chroboczek, H. and J. H. Cherry. *J. Mol. Biol.* 19: 28 (1966).
Ellem, K. A. O. *J. Mol. Biol.* 20: 1 (1966).
Hayashi, M., M. N. Hayashi and S. Spiegelman. *Proc. Natl. Acad. Sci. U.S.* 50: 664 (1963).

EXPERIMENT

20 POLYACRYLAMIDE GEL ELECTROPHORESIS OF NUCLEIC ACID

A. OBJECTIVE

To study the quantity and quality of nucleic acids it is essential to fractionate the classes of nucleic acids on the basis of size or molecular weight. One of the easiest and most useful techniques for this purpose is polyacrylamide gel electrophoresis. The objective of this experiment is for the student to learn the gel electrophoresis method of nucleic acid separation.

B. THEORY

Charged ions move in an electric field as a function of their total dimension and their net charge, as well as of the nature of the solution and current applied. Since the phosphate groups of nucleic acids are completely ionized at neutral pH, such molecules will move in an electric field toward the anode. Loening (1) has adapted the polyacrylamide gel electrophoresis technique to nucleic acids. The relative electrophoretic mobility of nucleic acid in polyacrylamide gel is inversely proportional to the log of the molecular weight. The mobility of the nucleic acid is also a function of the gel concentration (porosity). Figure 48 illustrates the concept.

FIGURE 48. The relation of molecular weights of RNA to the electrophoretic mobility in gels.

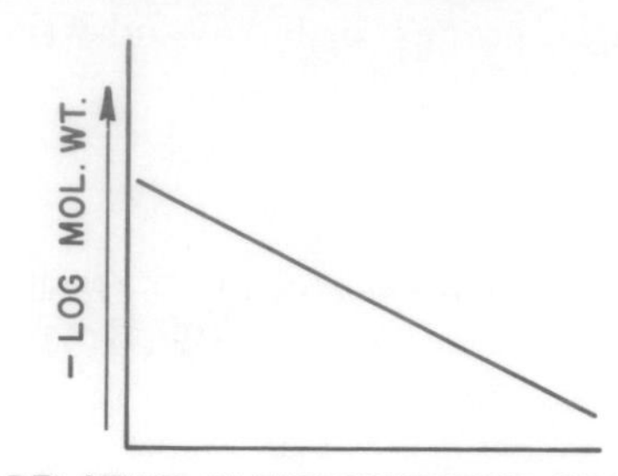

FIGURE 49. Separation of plant and bacterial ribosomal RNAs by polyacrylamide gel electrophoresis for 3.5 hr.

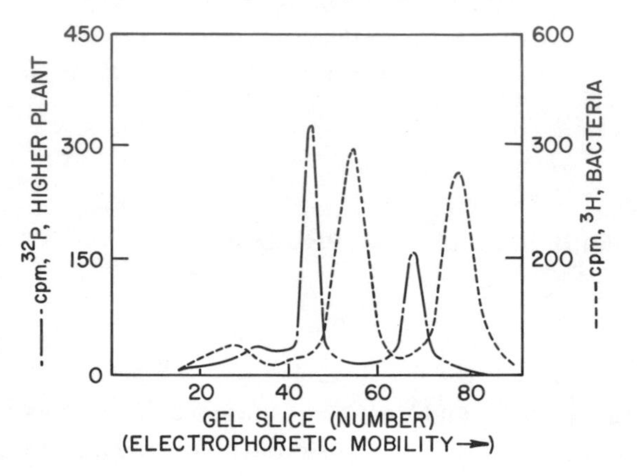

As shown in Figure 49, polyacrylamide gel electrophoresis is an excellent technique for the separation of RNA. The RNA from plant ribosomes (80S) is separated from the RNA of bacterial ribosomes (70S).

C. EQUIPMENT

Plexiglas electrophoresis apparatus with tubes (12 cm long, 0.6 cm in internal diameter), buffer chambers, and electrodes (a number of commercial units are available)
Beckman Duostat—regulated DC power supply
Chromoscan (Joyce Loeble & Co.)
Gel slicer (Mickle Lab. Eng. Co.)

D. SOLUTIONS

1. 15% Acylamide, 0.75% *N.N′*-methylene bisacrylamide; store in the refrigerator

2. 3E Buffer: 0.12 *M* Tris, 0.06 *M* sodium acetate, 0.003 *M* sodium EDTA; adjusted to pH 7.2 with glacial acetic acid

3. E Buffer (running buffer), upper and lower, total 4 liters:* 0.04 *M* Tris, 0.02 *M* sodium acetate, 0.001 *M* sodium EDTA; adjusted to pH 7.2 with glacial acetic acid; (running buffer can be reused up to six times, but pH should be checked each time)

4. TEMED (*N,N,N',N'*-tetramethylethylenediamine); store in the refrigerator

5. 10% Ammonium persulfate; must be freshly prepared

E. PURIFICATION OF CHEMICALS

1. Acrylamide

Dissolve 70 g in 1 liter of chloroform at 50 C. Filter under suction while hot. Chill to −15 C and recover crystals by filtration in the cold. Then wash on filter with cold chloroform or cold heptane.

2. Bisacrylamide

Dissolve 10 g in 1 liter of acetone at 40–50 C and filter while hot. Cool slowly to −15 C. Collect the crystals by centrifugation or by filtration. Then wash the crystals or precipitate with cold acetone.

3. Sodium Dodecyl Sulfate

Dissolve 20 g in 95% alcohol at 50 C and then filter the solution under suction. Store the filtrate at 4 C overnight and recover the precipitate by filtration under suction.

F. EXPERIMENTAL PROCEDURE

Purified RNA obtained in Experiment 19 or a radioactive sample provided by the instructor will be used for polyacrylamide gel electrophoresis. Detailed steps for gel electrophoresis of RNA follows.

1. Cover bottom of each gel tube with dialysis membrane (presoaked in water) and hold in place with a rubber ring cut from tubing.

2. Prepare a 2.4% gel solution, or other concentration as required (see Table 10).

3. Remove air from gel under vacuum in a desiccator.

4. Add 0.02 ml TEMED followed by 0.2 ml of 10% ammonium persulfate to the gel solution.

5. Swirl gel solution and carefully pipet it into gel tubes. Remove any excess gel solution with a dispo-pipet, adjusting the height of the

* Some investigators employ running buffer which contains 0.2% sodium lauryl sulfate.

TABLE 10

Additives	*Gel concentration* 2.4%	2.6%	2.7%	4.5%
Gel solution (solution 1), ml	4.00	4.33	4.50	7.50
Water, ml	12.45	12.12	11.95	8.95
3E Buffer, ml	8.33	8.33	8.33	8.33

solution to about 7 cm. If any gel is lost because of leaks from the lower end of tube, add extra gel solution to bring back to volume.

6. Allow gels to polymerize for about 20 min or longer. Upon polymerization, a straight line may be noted below the meniscus; it marks the upper surface of the polymerized gel.

7. Carefully pipet E buffer to fill the tops of the gel tubes. Pour E buffer into upper and lower buffer chambers. Enough buffer should be present to cover the top of the gels and for the lower ends to be submerged.

8. Prerun gels at room temperature for 30 min with power set at 5 mamp and 50 volts per tube.

9. Prepare the nucleic acid sample in about 50 μl (25–100 μl, depending on concentration of sample) of buffer made 5% with sucrose (sufficient to add 3–4 crystals of sucrose to 50-μl sample). A sample containing about 1 OD unit (50 μg) of RNA should be carefully layered on the gel surface with a dispo-pipet while current is on. This may be accomplished by carefully placing the tip of pipet along the inner wall of the gel tube and gently forcing the sample out with the aid of a suction bulb. As the sample will be denser than the buffer, it will settle on top of the gel.

10. Electrophoresis is carried out for 1.5 to 3.5 hr at 5 mamp per gel, depending on the nucleic acid sample. Soluble RNA is run off the gel by 3.5 hr, but better separation of the rRNA components is obtained in that time.

11. At the termination of the electrophoresis period, turn off the power supply and remove the gel tubes from the electrophoresis apparatus. Then remove and discard the dialysis tubing, and with the use of a large rubber bulb force the gel into a petri dish of water. Mark the dish opposite the upper end of the gel. Leave for about 30 min.

12. Suck the gel into a 5-ml modified pipet (mouth end removed) by means of a small rubber bulb at the pointed end. Transfer the gel to a special cuvette for use in a Chromoscan, and record the OD (wavelength near 260 mμ) profile of the gel.

13. Using the 5-ml pipet, transfer the gel to a hand-made aluminum foil tray of approximately the same size as the gel. Place small cylinders of Styrofoam at the ends of the gel to hold it in place. Adjust the gel to a standard length (approximately 6.8 cm) and then freeze with powdered dry ice.

14. Use a Gel Slicer to cut the frozen gel into 0.7-mm slices.

15. As the gel is sliced, remove the slices in order with forceps and place them on a long strip of filter paper. Dry the slices under an infrared lamp. Cut the filter paper into squares, each one bearing one dried slice, and place each one in a vial containing scintillation fluid. Then determine the amount of radioactivity in a scintillation spectrometer.

G. TREATMENT OF DATA

Prepare a short written report of your data (including figures) and discuss fractionation on gels in comparison to other techniques.

REFERENCE

1. Loening, U. E. *Biochem. J.* 102: 251 (1967).

General References

Bishop, D. H. L., J. R. Claybrook and S. Spiegelman. *J. Mol. Biol.* 26: 373 (1967).

Davis, B. J. *Ann. N.Y. Acad. Sci.* 121: 404 (1964).

EXPERIMENT 21 SEPARATION OF NUCLEIC ACIDS BY SUCROSE GRADIENT CENTRIFUGATION

A. OBJECTIVE

One of the most popular techniques for the fractionation of nucleic acids has been sucrose gradient centrifugation. RNA species sediment in a linear sucrose gradient when a centrifugal force is applied to the gradient. The rate of sedimentation of nucleic acid species in the sucrose gradient is related to their molecular size, the heavier species sedimenting more rapidly than the ligher species.

This experiment is designed to give the student a practical demonstration of the sucrose gradient technique of nucleic acid fractionation.

FIGURE 50. A system to produce three identical linear gradients.

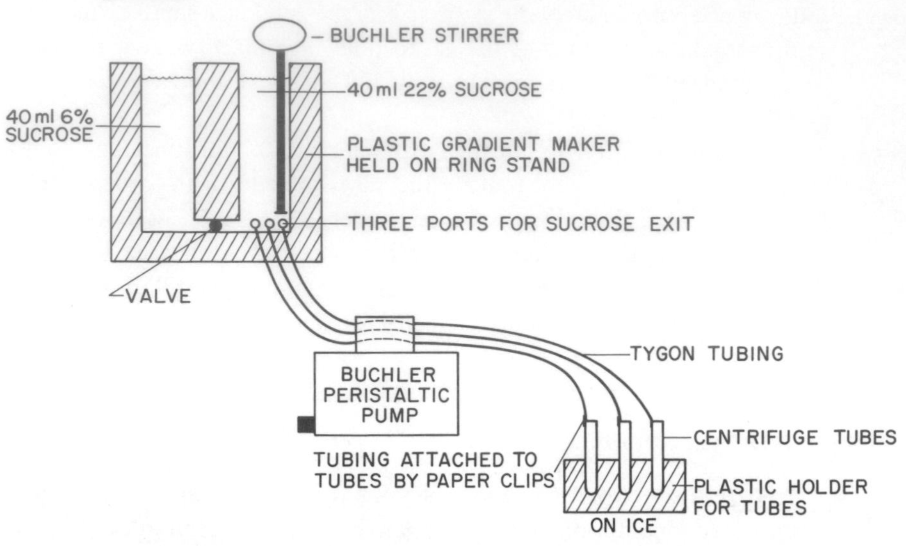

B. EQUIPMENT AND SUPPLIES

Plant ribonucleic acid
Sucrose solutions (6% and 22%) in 0.01 *M* Tris, pH 7.8, 10^{-3} *M* EDTA, and 0.05% Dupanol
Spinco model L centrifuge
SW 25 Rotor and cellulose-nitrate centrifuge tubes
Sucrose gradient maker
Pump for gradient maker
Gradient fractionator

C. EXPERIMENTAL PROCEDURE

Use samples of purified plant nucleic acid from Experiment 19. If you do not have an RNA sample, obtain one from the instructor or from another student in the course. The first step of the experiment is to make three sucrose gradients, using the gradient maker (ask the instructor to demonstrate its use). Three linear sucrose gradients ranging from 22% (bottom) to 6% (top) are made from 40 ml of each of the two sucrose solutions (RNase-free) which also contain 0.01 *M* Tris buffer, pH 7.8, 10^{-3} *M* EDTA, and 0.05% Dupanol.

The apparatus used to form the gradients is diagrammatically illustrated in Figure 50.

Layer a sample of nucleic acid, which contains about 10 OD_{260} units in a volume less than 1 ml, over the top of each sucrose gradient. Place the tubes in the SW-25 rotor and centrifuge at 25,000 rpm for 20 hr. *Make sure the brake is off and set the time to "hold."* After the allocated time, turn the timer on the centrifuge from hold to zero and, when it coasts to a stop, remove the tubes. Fractionate the gradients, using the gradient collection method (the same as used in Experiment 16). Manually collect fractions (10 drops/fraction) from the bottom of the tube in small test tubes. Add 2 ml of water to the tubes and read the absorbancy at 260 mμ. If the nucleic acid is radioactive, after reading absorbancy add 50 μg of DNA (0.05 ml of 1 mg/ml) and sufficient TCA (0.15 ml of 55%) to give a final concentration of 10%. The precipitated fractions can then be collected on glass fiber (Whatman GF/A) filters, dried, and counted in the liquid scintillation spectrometer.

D. TREATMENT OF DATA

Collect the data and plot the results as fraction number versus OD and cpm in each fraction. Write a short report summarizing the results.

GENERAL REFERENCES

Asano, K. *J. Mol. Biol.* 14: 71 (1965).
Barber, R. and M. Noble. *Biochim. Biophys. Acta* 123: 205 (1966).
Ellem, K. A. O. *J. Mol. Biol.* 20: 283 (1966).
Gros, F., W. Gilbert, H. H. Hiatt, G. Attardi, P. F. Spahr and J. D. Watson. *Cold Spring Harb. Symp. Quant. Biol.* 26: 111 (1961).
Hayes, D. H., F. Hayes and M. F. Gruerin. *J. Mol. Biol.* 18: 499 (1966).
Marcot-Queiroz, J. and R. Monier. *J. Mol. Biol.* 14: 490 (1965).
Moore, P. B. *J. Mol. Biol.* 22: 145 (1966).

EXPERIMENT 22 CHROMATIN–RNA POLYMERASE ASSAY

A. OBJECTIVE

Chromatin may be defined as interphase chromosomes. Chromatin material is heterogeneously composed of DNA, RNA, histones, and other basic proteins, including RNA polymerase and various factors involved in transcription. The objective of this experiment is to learn to isolate chromatin and to assay the RNA polymerase activity associated with it.

B. SUPPLIES AND MATERIALS

Etiolated soybean seedlings (4 days old)
Vir-Tis homogenizer
Dounce tissue grinder
Ultracentrifuge
Nucleoside triphosphates
^{3}H-UTP (12.1 mCi/μmole)
Glass fiber filters (Whatman GF/A)
Scintillation spectrometer

C. SOLUTIONS

Grinding Medium	*Wash Medium 1*	*Sucrose Solution*
0.25 *M* Sucrose	0.25 *M* Sucrose	2 *M* Sucrose
0.05 *M* Tris, pH 8.0	0.01 *M* Tris, pH 8.0	0.01 *M* Tris, pH 8.0
0.001 *M* $MgCl_2$	0.01 *M* 2-Mercapto-ethanol	0.01 *M* 2-Mercapto-ethanol
0.015 *M* 2-Mercapto-ethanol		

D. EXPERIMENTAL PROCEDURE

1. Chromatin Isolation

The procedure for chromatin extraction is exactly the same as the one given in Experiment 18, except that the final chromatin pellet is suspended in 1–2 ml of 0.01 *M* Tris buffer, pH 8.0.

2. RNA Polymerase Activity

The incorporation of ^{3}H-UTP into trichloroacetic acid–insoluble material is employed to measure RNA polymerase activity. The standard reaction mixture contains the following in 0.2 ml:

Chromatin equivalent to 2–3 μg of DNA in 0.05 ml of Tris
0.1 μmole GTP
0.1 μ mole ATP
0.1 μmole CTP
0.005 μmole UTP
7.5 μCi ^{3}H-UTP
0.5 μmole $MgCl_2$
0.125 μmole $MnCl_2$
0.5 μmole Cleland's reagent
10 μmoles Tris-HCl (pH 8.0)

The reaction is initiated by the addition of chromatin to a standard reaction mixture at 37 C. The reaction mixture is incubated for 15 min and then stopped by the addition of 4 ml of ice-cold 10% trichloroacetic acid containing 10^{-3} *M* inorganic pyrophosphate. Allow the precipitate to form for 15 min in the cold. Collect the precipitate on glass fiber filters (Whatman GF/A) and wash with 20 ml of cold 5% trichloroacetic acid, using a Millipore filter apparatus. Dry the filters under infrared lamps and then place the filters in vials containing scintillation fluid. Determine the radioactivity of each sample with a scintillation spectrometer.

Measure the chromatin DNA by the diphenylamine procedure of Burton (see Experiment 11) after hydrolysis in 0.5 *M* perchloric acid for 45 min at 70 C.

E. TREATMENT OF RESULTS

Calculate the pmoles (10^{-12} moles) of ^{3}H-UMP incorporated per 100 μg of DNA. Present the results as an abstract for *Plant Physiology*.

GENERAL REFERENCES

Hardin, J. W., T. J. O'Brien and J. H. Cherry. *Biochim. Biophys. Acta* 224: 667 (1970).

Holm, R. E., T. J. O'Brien, J. L. Key and J. H. Cherry. *Plant Physiol.* 45: 41 (1970).

Huang, R. C. C. and J. Bonner. *Proc. Natl. Acad. Sci. U.S.* 48: 1216 (1962).

Leffler, H. R., T. J. O'Brien, D. V. Glover and J. H. Cherry. *Biochem. Biophys. Res. Commun.* 38: 224 (1970).

O'Brien, T. J., B. C. Jarvis, J. H. Cherry and J. B. Hanson. *Biochim. Biophys. Acta* 169: 35 (1968).

PART

9

PROTEIN SYNTHESIS

INTRODUCTION

Practically all the nucleic acids are synthesized in the nucleus (ribosomal RNA in nucleolus), whereas virtually all proteins are produced in the cytoplasm. Thus two of the major questions facing biochemists today are: What are the mechanisms that specifically regulate the synthesis of a protein molecule, and how does the newly synthesized protein molecule get from the site of synthesis to its site of action? Because a number of components (tRNAs, ribosomes, mRNAs, enzymes, factors, etc.) are essential for protein synthesis, it is difficult to comprehend the significance of cellular spatial requirements before some understanding of the control mechanism for protein synthesis is gained. This part deals with the various components necessary to support *in vitro* protein synthesis (1, 2).

BACKGROUND

The first chemical indication that ribonucleoproteins played a direct role in the synthesis of proteins came in 1950 from "tracer" experiments in which animals were given labeled amino acids. When the cells of the labeled tissue were excised, homogenized, and then fractionated, the highest concentration of labeled amino acids was found in the microsomal fraction. This fraction subsequently has been shown to contain ribosome (and polyribosome) particles attached to membrane fragments. From careful kinetic studies on the rate of incorporation of labeled amino acids into different proteins of animal cells, it seemed very likely that the newly synthesized polypeptides attached to ribosomes were precursors to the enzymes found in the cells. Furthermore, it was found that RNase stopped protein synthesis in isolated subcellular fractions. The experiments made on whole animals and in isolated cells were soon supplemented by studies of cell-free systems containing RNA, which would incorporate amino acids into protein in the test tube. Subsequently, the main part of these studies on *in vitro* protein synthesis was accomplished with bacterial systems.

INCORPORATION OF AMINO ACIDS INTO PROTEIN

A prerequisite to the synthesis of protein is the acylation of amino acids to the proper transfer RNA species. The first step in protein synthesis is the activation of the amino acids with ATP by specific aminoacyl-tRNA synthetases (Figure 51).

FIGURE 51. Steps of aminoacylation of transfer RNA.

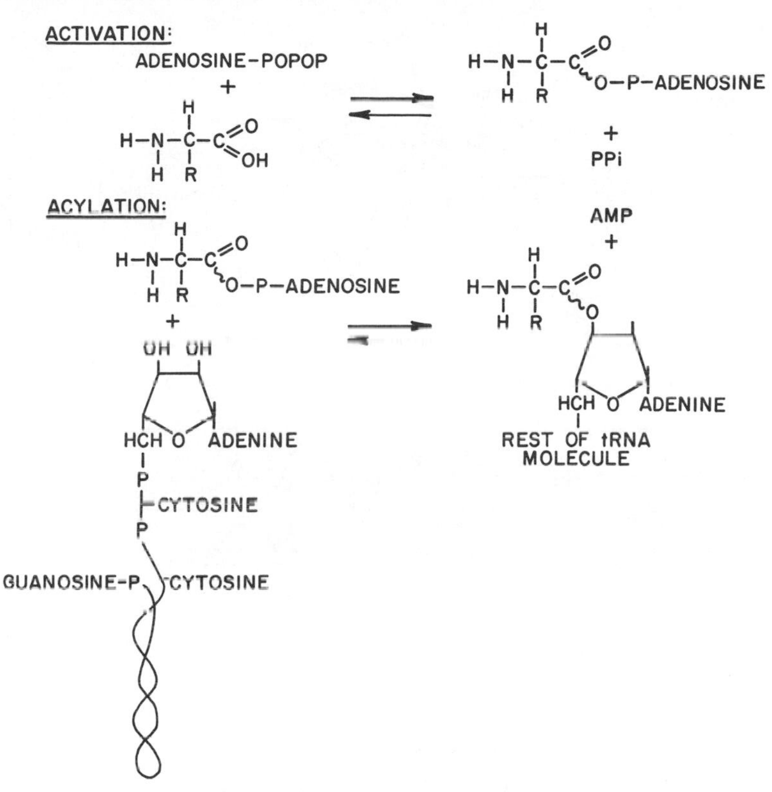

This reaction is followed by the attachment of the activated amino acids to specific tRNA molecules to form aminoacyl-tRNA. The acylation reaction is catalyzed by the same synthetase that is involved in amino acid activation. Although only a few aminoacyl synthetases have been purified, it is thought by many that only one specific enzyme is necessary for each amino acid. Recent information shows that there are at least three leucyl-tRNA synthetases in soybean cotyledons.

The final step of protein synthesis requires ribosomes, mRNA, GTP, and a number of factors. Even though a great deal of work has been done on protein synthesis in plants, most of our knowledge of this mechanism is derived from experiments using 70S microbial ribosomes, particularly those of *E. coli.* A diagrammatic representation of the present knowledge is indicated in Figure 52.

1. Mechanism of Protein Synthesis

(a) Binding of mRNA to 30S subunit.
(b) Attachment of Fmet-tRNA to 30S-mRNA complex.
(c) Addition of 50S subunit.

FIGURE 52. Proposed sequential reactions of protein synthesis.

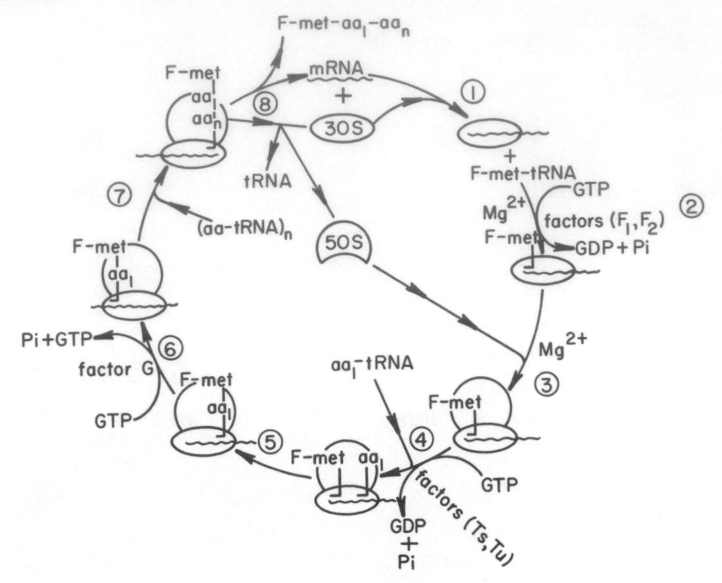

(d) Attachment of second incoming aa-tRNA.
(e) Peptide bond formation.
(f) Peptidyl translocation.
(g) Continuation of steps d, e, and f.
(h) Completion of translation and termination of protein.

TRANSFER RNA

Transfer RNA (often called soluble RNA) contains a great deal of helical (tertiary) structure. Most tRNA molecules are composed of about 80 nucleotides and have a molecular weight of 2.5×10^4 with a corresponding sedimentation coefficient of 4S. Transfer RNA is usually represented as a cloverleaf structure as indicated in Figure 53.

The tRNA molecule is composed of an anticodon loop (III), a minor loop (II), two other major loops (I and IV), and the stem, which contains an ACC terminus at the 3′-OH end. The stem is thought to be the site of enzyme recognition, while the anticodon loop contains the nucleotide triplet to bind the complementary codon triplet of the mRNA. At the moment, it is not known what part of the tRNA molecule is recognized by the ribosome. The major function of tRNA is to provide a selective and precise manner of placing an amino acid into a growing peptide chain.

FIGURE 53. Generalized structure of transfer RNA (1).

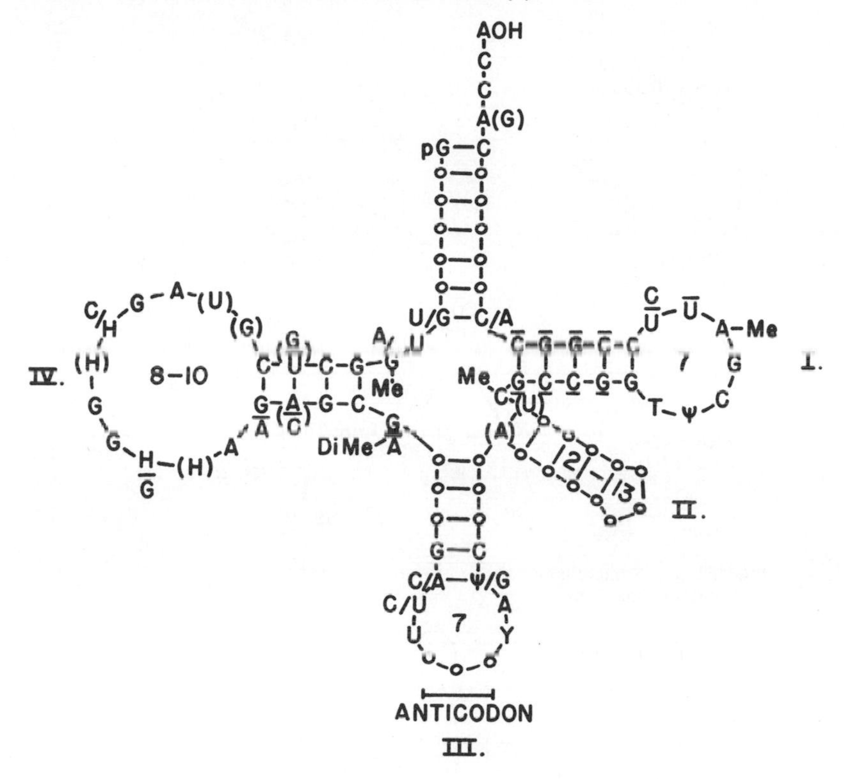

RIBOSOMES FROM PROCARYOTIC AND EUCARYOTIC ORGANISMS

Ribosomes generally contain 60–65% RNA and 35–40% protein. Bacterial ribosomes and ribosomes from mitochondria and chloroplasts have sedimentation coefficients of 70S, while those from the cytoplasm of eucaryotic organisms have a coefficient of 80S. Some general properties of *E. coli* ribosomes are indicated in Figure 54.

Ribosomes from eucaryotic organisms contain subunits and RNA species as indicated in Figure 55.

GENETIC CODE

The genetic code is made from an alphabet containing four letters (the four nucleotide bases) and composed of three-letter code words. Thus there are 4^3 or 64 possible code words which specifically allow the 20 amino acids to be placed in polypeptides in a precise sequence. These code words for each amino acid were assigned by employing synthetic mRNAs composed of different quantities of the four ribonucleotides. In 1961 Nirenberg (3) and his associates discovered that UUU (poly U) codes for phenylalanine. This was the first code word known,

FIGURE 54. Characteristics of bacterial (70S) ribosomes.

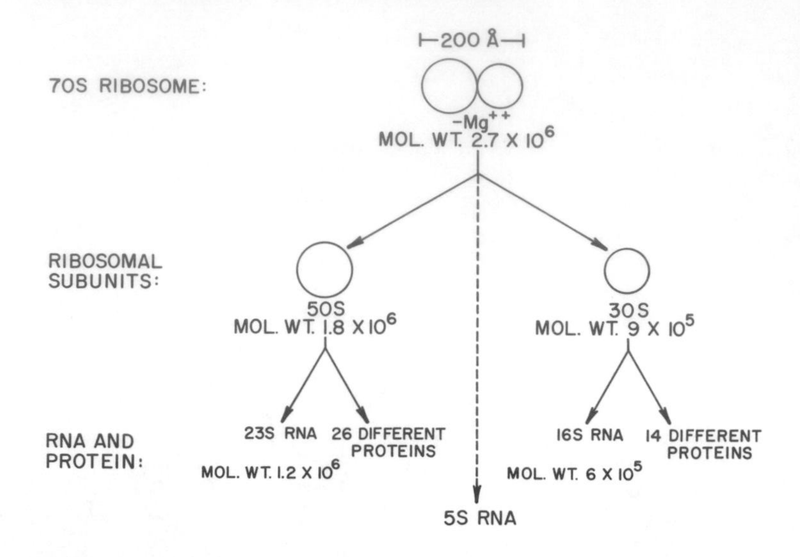

FIGURE 55. Some characteristics of ribosomes (80S) from higher organisms.

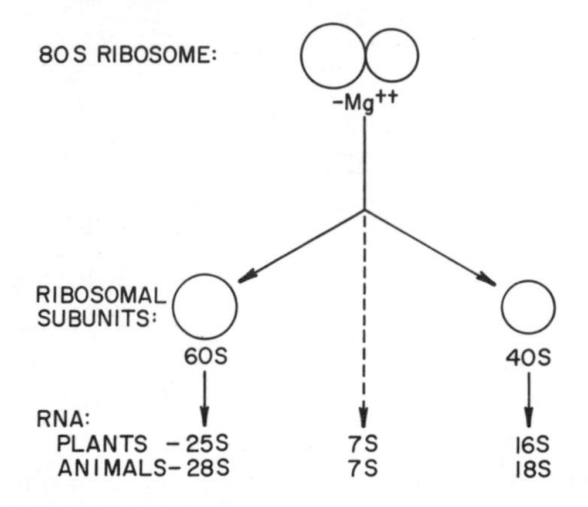

and similar studies led to the complete elucidation of the genetic code. An mRNA of known nucleotide composition (but unknown sequence) was supplied to monoribosomes, and the incorporation of all amino acids (1 labeled, 19 unlabeled) was separately tested. This was repeated for a large number of synthetic mRNAs. The data thus collected were analyzed for best fit, and the code words for the amino acids were assigned. The assignments of the 64 code words are presented in Table 11. The evidence used to produce Table 11 comes mainly from studies on *E. coli*. It is likely that the genetic code in other organisms is either very similar to or identical with that shown here.

TABLE 11. THE GENETIC CODE

1st ↓	2nd → U	C	A	G	3rd ↓
	PHE	SER	TYR	CYS	U
	PHE	SER	TYR	CYS	C
U					
	LEU	SER	Ochre	Nonsense	A
	LEU	SER	Amber	TRP	G
	LEU	PRO	HIS	ARG	U
	LEU	PRO	HIS	ARG	C
C					
	LEU	PRO	GLN	ARG	A
	LEU	PRO	GLN	ARG	G
	ILEU	THR	ASN	SER	U
	ILEU	THR	ASN	SER	C
A					
	ILEU	THR	LYS	ARG	A
	MET	THR	LYS	ARG	G
	VAL	ALA	ASP	GLY	U
	VAL	ALA	ASP	GLY	C
G					
	VAL	ALA	GLU	GLY	A
	VAL	ALA	GLU	GLY	G

In bacteria the UAA and UAG triplets are termination codons, while the UAG and UGA triplets are considered termination codons in higher organisms.

REFERENCES

1. Jachymczyk, W. J. and J. H. Cherry. *Biochem. Biophys. Acta* 157: 368 (1968).
2. Marcus, A., B. Luginbill and J. Feeley. *Proc. Natl. Acad. Sci. U.S.* 59: 1243 (1968).
3. Nirenberg, M. W. and J. H. Matthaei. *Proc. Natl. Acad. Sci. U.S.* 47: 1588 (1961).

General References

Bhargava, P. M., T. Pallaiah and E. Premkumar. *J. Theor. Biol.* 29: 447 (1970).

Boulter, D. *Ann. Rev. Plant Physiol.* 21: 91 (1970).

Cold Spring Harbor Laboratory. *Cold Spring Harb. Symp. Quant. Biol.* 14: 855 (1969).

Huystee, van R. B., W. J. Jachymczyk, C. F. Tester and J. H. Cherry. *J. Biol. Chem.* 243: 2315 (1968).

Kanabus, J. and J. H. Cherry. *Proc. Natl. Acad. Sci. U.S.* 68: 873 (1971).

Mans, R. J. and G. D. Novelli, *Biochem. Biophys. Acta* 50: 287 (1961).

Marcus, A. and J. Feeley. *J. Biol. Chem.* 240: 1675 (1965).
Marcus, A. and J. Feeley. *Proc. Natl. Acad. Sci. U.S.* 56: 1770 (1966).
Nirenberg, M. W., O. W. Jones, P. Leder, B. V. C. Clark, W. S. Sly and S. Peska. *Cold Spring Harb. Symp. Quant. Biol.* 28: 549 (1963).
Phillips, G. R. *Nature* 223: 374 (1969).

EXPERIMENT 23 ISOLATION OF TRANSFER RNA

A. OBJECTIVE

The first step in protein synthesis is the activation of the amino acid and the subsequent transfer to specific tRNA molecules. The objective of this experiment is to learn the techniques of extracting tRNA from plant tissue and the requirements for acylation of tRNA with a labeled amino acid.

B. EQUIPMENT AND SUPPLIES

Plant tissue (soybean cotyledons)
Washed phenol
Buffers: (1) 0.003 *M* Tris-succinate, pH 7.4, 0.03 *M* magnesium acetate, 0.003 *M* EDTA, and 0.005 *M* 2-mercaptoethanol; (2) TMP—0.1 *M* Tris-HCl, pH 7.8, 0.05 *M* $MgCl_2$, and 2% PVP
2 *M* Potassium acetate
95% Ethanol
Cheesecloth
Homogenizer
Dialysis tubing
U.V. spectrophotometer
^{3}H-L-Leucine, 2 Ci/mmole
0.05 *M* ATP, pH 7.8
Whatman No. 3 paper discs (2.5 cm)
Insect pins
Various TCA solutions (see Experiment 7)
Refrigerated centrifuge
Magnetic stirrers

C. EXPERIMENTAL PROCEDURE

1. Extraction of tRNA

Transfer RNA will be extracted from 20 g of 4-day-old etiolated soybean cotyledons (or some other tissue of your choice) by a method adapted from Zubay (1). Grind the tissue in an Omni-Mixer (or a

TABLE 12

	Reaction number				
Additives, ml	*1*	*2*	*3*	*4*	*5*
TMP (0.1 *M* Tris-HCl, pH 7.8, 0.05 *M* $MgCl_2$, 2% PVP)	0.1	0.1	0.1	0.1	0.1
ATP (0.05 *M*, pH 7.8)	0.1	0.1	0.1	0.1	0.1
tRNA, 1 mg/ml	0	0.05	0.1	0.2	0.5
^{3}H-L-Leucine, 2 Ci/mmole	0.001	0.001	0.001	0.001	0.001
H_2O	0.7	0.65	0.6	0.5	0.2
Synthetase fraction, 5 mg/ml	0.1	0.1	0.1	0.1	0.1

similar homogenizer) in 10 ml of a medium of 0.03 *M* magnesium acetate, 0.003 *M* Tris-succinate, pH 7.4, 0.003 *M* EDTA and 0.005 *M* 2-mercaptoethanol, and 15 ml of phenol. Stir the slurry with the homogenizer for about 30 min and then centrifuge at 20,000 × *g* for 15 min. Remove the aqueous phase with a hypodermic needle and syringe and then add solid potassium acetate (to make 0.2 *M*) followed by 2 volumes of cold 95% ethanol. Place the mixture on a magnetic stirrer in the cold and stir for 1–2 hr. Collect the resulting precipitate at 10,000 × *g* for 15 min. Suspend the precipitate in a small volume of 2 *M* potassium acetate by stirring in the cold for 2–4 hr. Make sure that all particles are broken. Remove the 2 *M* potassium acetate–insoluble material by centrifugation at 10,000 × *g* for 15 min. (If the aqueous phase is cloudy, reextract it with phenol.) Then add 2 volumes of cold 95% ethanol to the aqueous supernatant and allow the tRNA to precipitate in the cold (−15 C best) for 2 hr or longer. Collect the precipitate by centrifugation at 10,000 × *g* for 15 min. Dissolve the pellet in a small amount of water (5 ml or less) and dialyze overnight against water containing 0.005 *M* 2-mercaptoethanol. Determine the tRNA content of the dialyzed material by reading the absorption of a diluted (1/50) sample at 260 mμ and an absorption spectrum from 220–300 mμ on the spectrophotometer. Usually one A_{260} unit is equal to 50 μg of tRNA. If the tRNA concentration is greater than 1 mg/ml, dilute to that concentration and freeze for further studies.

2. Aminoacylation Assay

The acylation of the tRNA preparation will be done with an enzyme obtained from Experiment 24. The general assay conditions are essentially the same as those given in Experiment 7 for leucyl-tRNA synthetase assay. The components of the reaction (total volume of 1 ml) are indicated in Table 12.

The arrangement of the filter discs, monitoring the amount of aminoacylation with time, processing of the labeled disks, assay of radioactivity, etc., are exactly as outlined in Experiment 7.

D. TREATMENT OF DATA

Present the results (the U.V. absorption spectrum and the leucyl-acceptor activity of the isolated tRNA) in the form of figures or tables. Also calculate the percentage of leucine-specific tRNA in the preparation.

REFERENCE

1. Zubay, G. *J. Mol. Biol.* 4: 347 (1962).

General References

Anderson, M. B. and J. H. Cherry. *Proc. Natl. Acad. Sci. U.S.* 62: 202 (1969).
Stent, G. S. *Science* 144: 816 (1964).

EXPERIMENT 24 ISOLATION OF AMINOACYL-tRNA SYNTHETASE

A. OBJECTIVE

The rate of protein synthesis is dependent on a number of components, including the levels of aminoacyl-tRNA synthetases for each amino acid. In order to assay *in vitro* amino acid incorporation, therefore, it is important that a good synthetase preparation which contains acylation activities for all amino acids, if possible, be available. The objective of this experiment is to provide a fairly simple procedure for the isolation of a synthetase preparation.

B. EQUIPMENT AND SUPPLIES

Plant tissue (soybean cotyledons)
Polyclar-AT (insoluble polyvinyl pyrrolidine)
Various buffers: (1) 0.1 *M* Tris, pH 7.9, 0.04 *M* KCl, 0.04 *M* $MgCl_2$, and 0.01 *M* 2-mercaptoethanol; (2) 0.01 *M* sodium phosphate, pH 6.0; (3) TMP: 0.1 *M* Tris-HCl, pH 7.8, 0.05 *M* $MgCl_2$, and 2% PVP
Cheesecloth
Dialysis tubing

Homogenizer
U.V. spectrophotometer
Solid $(NH_4)_2SO_4$ (enzyme grade)
CM-cellulose
Glass column
^{3}H-L-Leucine (2 Ci/mmole, or higher)
0.05 *M* ATP, pH 7.8
Refrigerated centrifuge
Insect pins
tRNA (Experiment 23)
Whatman No. 3 filter paper discs (2.5 cm)
Magnetic stirrers
Various TCA solutions

C. EXPERIMENTAL PROCEDURE

1. Enzyme Preparation

Aminoacyl-tRNA synthetases will be prepared essentially according to the method of Anderson and Cherry (1). Remove 10 g of soybean cotyledons from 4-day-old soybean seedlings and chill in ice. Grind the tissue by hand in a prechilled mortar and pestle in a solution (1 ml/g) containing 0.1 *M* Tris, pH 7.9, 0.04 *M* KCl, 0.04 *M* $MgCl_2$, and 0.01 *M* 2-mercaptoethanol. When the tissue is macerated, add 10 g of Polyclar-AT saturated with half-strength homogenizing medium. Mix the Polyclar-AT with the homogenate, and then filter the resulting slurry through cheesecloth by applying manual pressure. Centrifuge the filtrate in the ultracentrifuge at 75,000 × *g* for 1 hr. Dialyze the supernatant for 2–4 hr against 1–2 liters of cold half-strength homogenizing solution and again centrifuge for 1 hr at 75,000 × *g*. At this stage the solution may be frozen for further use. Subsequently, the solution is saturated to 70% with respect to $(NH_4)_2SO_4$ (see the chart in Appendix 1). Care should be taken to add the solid $(NH_4)_2SO_4$ slowly to ensure gradual precipitation of the enzyme protein. Collect the precipitate by centrifugation at 10,000 × *g* for 15 min and then dissolve it in 0.01 *M* sodium phosphate buffer, pH 6.0.

Pour into a glass column (2.1 × 40 cm) a slurry of CM-cellulose saturated with the phosphate buffer and pack to a height of 6 cm. Wash the column with 15–20 ml of phosphate buffer and then add the protein preparation to the column. Wash the protein fraction onto and through the column with the same buffer. Manually collect fractions containing 5 ml each and read the absorbance at 280 mμ. A single protein peak is eluted from the column in the first ten fractions. Pool

TABLE 13

Additives, ml	*Reaction number*				
	1	*2*	*3*	*4*	*5*
TMP (0.1 *M* Tris-HCl, pH 7.8, 0.05 *M* $MgCl_2$, 2% PVP)	0.1	0.1	0.1	0.1	0.1
ATP (0.05 *M*, pH 7.8)	0.1	0.1	0.1	0.1	0.1
tRNA (mg/ml) yeast or tRNA from Experiment 23	0.2	0.2	0.2	0.2	0.2
^{3}H-L-leucine (2 Ci/mmole or higher)	0.001	0.001	0.001	0.001	0.001
H_2O	0.6	0.59	0.55	0.5	0.1
Synthetase fraction (5 mg/ml)	0	0.01	0.05	0.1	0.5

the fractions containing the peak absorbancy at A_{280} and saturate to 70% with solid $(NH_4)_2SO_4$. Collect the precipitated protein by centrifugation and dissolve it in a small volume (2–3 ml) of half-strength homogenizing medium. Determine the protein content by the method of Warburg and Christian (Experiment 7). Subsequently dilute the enzyme preparation to contain 5 mg protein/ml and use as a source of synthetase.

2. Measurement of Enzyme Activity

Acylation of yeast tRNA or tRNA prepared in Experiment 23 will be employed to estimate enzyme activity. The assay conditions are described in Experiments 7 and 23. The protocol of the experiment is indicated in Table 13.

Run the reaction at 27 C and take samples at intervals of 3, 6, 9, 12, 18, 25, 30, 45, and 60 min. Remove 0.1 ml from each sample and place on a separate filter paper disc and subsequently assay for radioactivity as indicated in Experiment 7.

D. TREATMENT OF DATA

In a short report present the results on leucyl-tRNA synthetase activity as related to enzyme concentrations. Under what acylation conditions is it possible to observe differences in synthetases from different tissues?

REFERENCE

1. Anderson, M. B. and J. H. Cherry. *Proc. Natl. Acad. Sci. U.S.* 62: 202 (1969).

General References

Bick, M. D., H. Liebke, J. H. Cherry and B. L. Strehler. *Biochem. Biophys. Acta* 204: 175 (1970).

Cherry, J. H. and D. J. Osborne. *Biochem. Biophys. Res. Commun.* 40: 763 (1970).

Kanabus, J. and J. H. Cherry. *Proc. Natl. Acad. Sci. U.S.* 68: 873 (1971).

EXPERIMENT 25 ISOLATION OF PLANT RIBOSOMES ON A SUCROSE GRADIENT

A. OBJECTIVE

Polyriboses are composed of ribosomes attached to mRNA. In order to study *in vitro* protein synthesis without added mRNA, a ribosomal fraction containing polyribosomes is required. Furthermore, the relative amount of polyribosomes to monoribosomes often provides information on *in vivo* protein synthesis. The objective of this experiment is to allow the student to learn methods dealing with ribosome fractionation.

B. EQUIPMENT AND SUPPLIES

Plant tissue (3-day-old dark-germinated peanuts and oats); also dry peanut seeds
Glass homogenizer
Homogenizing medium
Various solutions: (1) 0.05 *M* Tris, pH 7.8, 0.01 *M* $MgCl_2$, 0.005 *M* 2-mercaptoethanol, 0.5% deoxycholate, 0.25 *M* sucrose; (2) 0.01 *M* Tris, pH 7.8, 0.015 *M* KCl, 0.1 *M* $MgCl_2$, 1.6 *M* sucrose; (3) 0.01 *M* Tris, pH 7.8, 0.015 *M* KCl, 0.01 *M* $MgCl_2$, 0.005 *M* 2-mercaptoethanol; (4) 12% sucrose containing 0.1 *M* Tris, pH 7.8, 0.001 *M* spermidine, 0.01 *M* $MgCl_2$; (5) 36% sucrose containing 0.1 *M* Tris, pH 7.8, 0.001 *M* spermidine, 0.01 *M* $MgCl_2$.
Spinco centrifuge
Sucrose gradient apparatus
ISCO gradient fractionator

C. EXPERIMENTAL PROCEDURE

1. Preparation of Ribosomes

Remove the cotyledons from four (8–10 g) peanut seedlings (3 days after planting). Pulverize the tissue in a small prechilled mortar and pestle (*keep cold*) in 11 ml of 0.25 *M* sucrose solution containing 0.05 *M* Tris buffer, pH 7.8, 0.01 *M* $MgCl_2$, 0.005 *M* 2-mercaptoethanol, and 0.5% deoxycholate until a uniform homogenate is obtained. Filter the homogenate through two layers of cheesecloth and then centrifuge the homogenate at 20,000 × *g* for 10 min to remove cellular debris. Filter the supernatant through Miracloth or glass wool. Ribosomes may be obtained from the supernatant solution by layering the sample (4.5 ml) over 4 ml of 1.6 *M* sucrose (containing 0.01 *M* Tris, pH 7.8, 0.015 *M* KCl, and 0.1 *M* $MgCl_2$) followed by centrifugation at 105,000 × *g* for 3 hr in the Ti-50 Spinco rotor (a longer time may be necessary to get higher yields). Wipe the tubes free of lipids, etc., and gently suspend the ribosomal pellets in about 1 ml of 0.01 *M* Tris buffer, pH 7.8, containing 0.015 *M* KCl, 0.01 *M* $MgCl_2$, and 0.005 *M* 2-mercaptoethanol. This ribosomal suspension will be used for sucrose gradient fractionation. Similar procedures will be utilized for preparing ribosomes from the embryos of germinated oat or wheat seedlings except that the glass homogenizer may be used.

2. Preparation of Sucrose Gradient

Make a linear sucrose gradient from 12% (top) to 36% (bottom) using the sucrose gradient apparatus described in Experiment 21. The method of preparing the sucrose gradient will be demonstrated in the laboratory, if desired. Both sucrose solutions will contain the following ingredients: 0.1 *M* Tris, pH 7.8, 0.001 *M* spermidine, 0.01 *M* $MgCl_2$.

Layer the suspended ribosomal preparations over the top of the sucrose gradients.

3. Gradient Centrifugation

Place the tubes containing the samples in the SW 25.1 rotor and centrifuge for 3 hr at 23,000 rpm. *Make sure the brake is off*!

4. Fractionation of the Sucrose Gradient

When the centrifugation is complete, gently remove the tubes from the rotor buckets and place in a plastic holder for the tubes.

If available, an ISCO fractionator (or other automated instrument) connected to a U.V. analyzer will be employed to determine the ribosomal profiles. Alternatively, fractions from the tube will be collected manually with an apparatus illustrated in Figure 56.

Carefully puncture the centrifuge tube and allow the sucrose solution to flow drop by drop into the test tubes positioned below the apparatus. Control the flow rate from the apparatus to allow careful counting of

FIGURE 56. A device to collect manually the contents of the sucrose gradient.

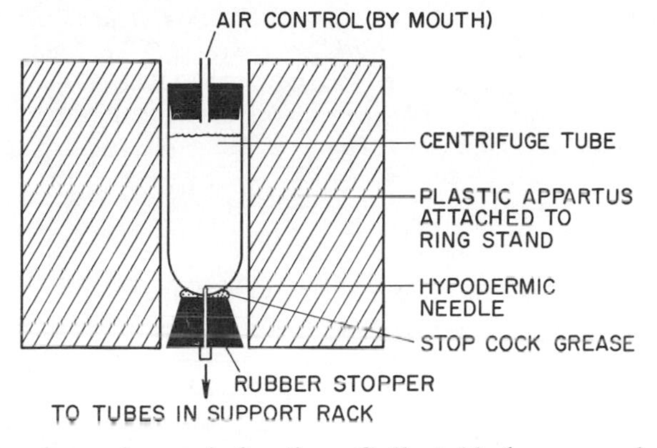

the drops for each fraction. Collect 10 drops per fraction. Then add 2 ml of H_2O and determine the U.V. absorbancy at 260 mμ using the spectrophotometer. See (1) for an example of polyribosome separation.

The basic objective of isolating polyribosomes is to obtain a high yield with little degradation (usually RNase degradation). A number of workers (2–4) have shown that diethyl pyrocarbonate (DEP) is an effective inhibitor of RNase when the chemical is added to the grinding medium. However, when high levels of DEP are used, the ribosomes are nonfunctional in regard to amino acid incorporation.

A review (5) of the isolation of microsomes, monoribosomes, and polyribosomes deals with a number of the above-mentioned procedures.

D. TREATMENT OF DATA

From your data prepare a short communication for *Biochimica et Biophysica Acta* showing ribosomal profiles from the three tissue samples.

REFERENCES

1. Dure, L. and L. Waters. *Science* 147: 410 (1965).
2. Anderson, J. M. and J. L. Key. *Plant Physiol.* 48: 801 (1971).
3. Travis, R. L., R. C. Huffaker and J. L. Key. *Plant Physiol.* 46: 800 (1970).
4. Weeks, D. P. and A. Marcus. *Plant Physiol.* 44: 1291 (1969).
5. Cherry, J. H. *Methods in Enzymology, Biomembranes.* Academic Press, New York. In press.

General References

Boyley, S. T. *J. Mol. Biol.* 8: 231 (1964).

Clark, M. F., R. E. F. Matthews and R. K. Ralph. *Biochim. Biophys. Acta* 91: 289 (1964).

Gierer, A. *J. Mol. Biol.* 6: 148 (1963).

Huystee, van R. B., W. Jachymczyk, C. F. Tester and J. H. Cherry. *J. Biol. Chem.* 243: 2315 (1968).
Jachymczyk, W. J. and J. H. Cherry. *Biochim. Biophys. Acta* 157: 368 (1968).
Leaver, C. J. and J. L. Key. *Proc. Natl. Acad. Sci. U.S.* 57: 1338 (1967).
Lin, C. Y. and J. L. Key. *Plant Cell Physiol.* 9: 553 (1968).
Marcus, A. and J. Feeley. *J. Biol. Chem.* 240: 1675 (1965).
Tester, C. F. and Dure, L. *Biochem. Biophys. Res. Commun.* 23: 287 (1966).

EXPERIMENT

26 POLYURIDYLIC ACID–DIRECTED PHENYLALANINE INCORPORATION

A. INTRODUCTION

Polyuridylic acid was the first synthetic polyribonucleotide discovered to have mRNA-like activity. None of its bases is normally hydrogen bonded in solution, and it binds well to free ribosomes. Under proper conditions it selects phenylalanyl-tRNA molecules exclusively, thereby forming a polypeptide chain containing only phenylalanine (polyphenylalanine). Thus we know that a codon for phenylalanine is composed of a group of three uridylic acid residues (UUU) since a three-base sequence constitutes a single codon. A schematic diagram of poly U–directed phenylalanine incorporation by plant ribosomes is indicated in Figure 57.

FIGURE 57. Polyuridylic acid–directed synthesis of polyphenylalanine on 80S ribosomes.

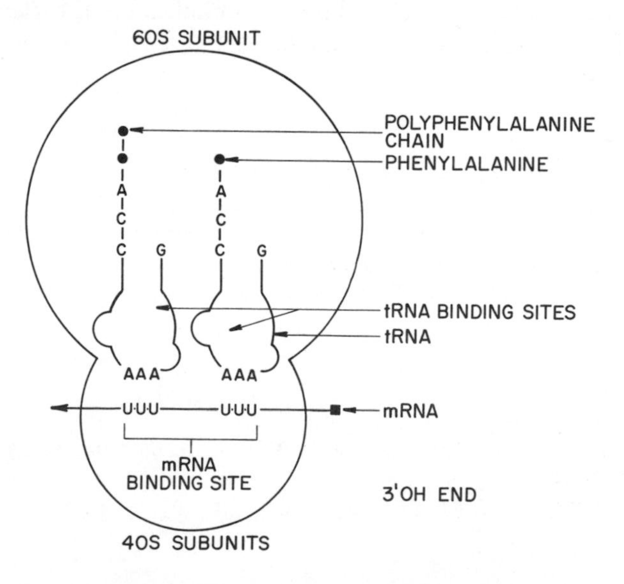

B. OBJECTIVES

The major objectives of this experiment are to learn how to isolate monoribosomes from plant tissue and to demonstrate poly U–directed phenylalanine incorporation.

C. EQUIPMENT AND SUPPLIES

Plant material (dry peanut seed)
Transfer RNA
Aminoacyl-tRNA Synthetase preparation (see Experiment 24)
ATP, GTP, creatine phosphate, creatine phosphate kinase
Ultracentrifuge
Whatman No. 3 filter paper discs (2.5 cm)
Insect pins
5% TCA
Various buffers: (1) 0.05 *M* Tris-succinate, pH 7.8, 0.01 *M* KCl, 0.01 *M* $MgCl_2$, 0.005 *M* 2-mercaptoethanol, 0.5% deoxycholate, 0.25 *M* sucrose; (2) 0.01 *M* Tris-succinate, pH 7.8, 0.01 *M* $MgCl_2$, 0.01 *M* KCl, 0.005 *M* 2-mercaptoethanol, and 1.6 *M* sucrose; (3) 0.01 *M* Tris-succinate, pH 7.8, 0.01 *M* $MgCl_2$, 0.01 *M* KCl, and 0.005 *M* 2-mercaptoethanol; (4) 0.5 *M* Tris, pH 7.8, 0.5 *M* KCl, 0.1 *M* $MgCl_2$, 0.02 *M* 2-mercaptoethanol
^{3}H-L-Phenylalanine (1 mCi/ml)
Polyuridylic acid
Scintillation spectrometer

D. EXPERIMENTAL PROCEDURE

1. Preparation of Ribosomes

Homogenize 10 peanut cotyledons (dry seed, skins removed) in a prechilled mortar and pestle in 10 ml (w/v is about 1:1) of a medium containing 0.25 *M* sucrose, 0.01 *M* $MgCl_2$, 0.05 *M* Tris-succinate buffer,* pH 7.8, 0.01 *M* KCl, 0.005 *M* 2-mercaptoethanol, and 0.5% deoxycholate. Strain the homogenate through cheesecloth and centrifuge for 15 min at 20,000 × *g*. Filter the supernatant through Miracloth, layer it over 4 ml of 1.8 *M* sucrose containing 0.01 *M* $MgCl_2$, 0.01 *M* KCl, 0.005 *M* 2-mercaptoethanol, and 0.01 *M* Tris-succinate, pH 7.8, in a centrifuge tube and then centrifuge at 105,000 × *g* for 3 hr in a Spinco ultracentrifuge (Ti-50 rotor). Discard the resulting supernatant. Resuspend the ribosomal pellets in 5 ml of a solution

* A Tris-succinate buffer is made by adjusting the pH of Tris to 7.8 with succinic acid.

TABLE 14

Additives, ml	*Reaction number*				
	1	*2*	*3*	*4*	*5*
1. Buffer (0.5 *M* Tris, pH 7.8, 0.5 *M* KCl, 0.1 *M* $MgCl_2$, 0.02 *M* 2-mercaptoethanol)	0.1	0.1	0.1	0.1	0.1
2. ATP (0.01 *M*)	0.1	0.1	0.1	—	0.1
3. GTP (0.005 *M*)	0.1	0.1	0.1	0.1	0.1
4. Creatine phosphate[a] (0.01 *M*)	—	0.1	0.1	0.1	0.1
5. Creatine phosphate[a] kinase	—	50 μg	50 μg	50 μg	50 μg
6. ^{3}H-L-Phenylalanine, 1 mCi/ml	0.01	0.01	0.01	0.01	0.01
7. Ribosomes, 3 mg/ml	0.1	0.1	—	0.1	0.1
8. Synthetase, 5 mg/ml	0.1	0.1	0.1	0.1	0.1
9. Poly U, 1 mg/ml	0.1	—	0.1	0.1	0.1
10. H_2O	0.39	0.39	0.39	0.39	0.29

[a] Creatine phosphate and creatine phosphate kinase are used as an ATP generator.

containing 0.01 *M* $MgCl_2$, 0.01 *M* KCl, 0.005 *M* 2-mercaptoethanol, and 0.01 *M* Tris-succinate buffer, pH 7.8. Clarify this solution by centrifuging for 10 min at 20,000 × *g*. Then dilute the supernatant to a final concentration of ribosomes* corresponding to 2–4 mg RNA/ml. Perform all operations in the cold (1–2 C).

2. *Conditions for Incorporation of ^{3}H-Phenylalanine*

The incubation medium is essentially that used by Marcus and Feeley (2, 3) and includes the following (in μmoles/ml of medium): Tris buffer, pH 7.8, 50; KCl, 50; $MgCl_2$, 10; 2-mercaptoethanol, 2; ATP, 1; GTP, 0.5; and creatine phosphate, 10. Creatine phosphate kinase (50 μg) and 5 μCi of labeled ^{3}H-L-phenylalanine are also supplied.

Usually 0.3 mg of ribosomes and 0.5 mg of aminoacyl-tRNA synthetase preparation are added to the 1-ml incubation mixture to initiate the reaction. Poly U is added to give a concentration of 100 μg/ml.

A list of the additives to be included in the reaction mixture is given in Table 14.

* An A_{260} value of 12 is equivalent to 1 mg of ribosomal protein and 0.5 mg of RNA (1) per ml.

3. Assay of Amino Acid Incorporation

Phenylalanine incorporation is initiated by adding to the tubes both ribosomes and synthetase at 30-sec intervals. The reaction mixture is incubated for 3, 6, 12, and 24 min and is stopped by placing 0.1-ml portions of the incubation medium on duplicate filter paper discs (see Experiments 7, 23, and 24). One set of the discs is treated in the same way as previously described (Experiment 7). However, the duplicate set is carried through an additional step. After incubation in 5% TCA, place these discs in 5% TCA and heat to 90 C for 10 min to discharge phenylalanyl-tRNA. The radioactivity remaining on the filter discs after this treatment will be in the form of peptides only. Complete the washing by incubating the discs in ethanol-ether. Dry and then place the discs in scintillation vials and determine the amount of radioactivity incorporated.

E. TREATMENT OF DATA

Write a short report showing rates of phenylalanine charging and incorporation into polypeptides. Calculate the percentage of the charged amino acid incorporated into a polypeptide as a function of time. Comment on the requirements of the incubation mixture of the system.

REFERENCES

1. Tester, C. F. and L. S. Dure, III. *Biochem. Biophys. Res. Commun.* 23: 287 (1966).
2. Marcus, A. and J. Feeley. *Proc. Natl. Acad. Sci. U.S.* 51: 1075 (1964).
3. Marcus, A. and J. Feeley. *J. Biol. Chem.* 240: 1675 (1965).

General References

Anderson, M. B. and J. H. Cherry. *Proc. Natl. Acad. Sci. U.S.* 62: 202 (1969).

Huystee, van R. B., W. Jachymczyk, C. F. Tester and J. H. Cherry. *J. Biol. Chem.* 243: 2315 (1968).

Jachymczyk, W. J. and J. H. Cherry. *Biochim. Biophys. Acta* 157: 368 (1968).

Keller, E. B. and P. C. Zamecnik. *J. Biol. Chem.* 221: 45 (1956).

Mans, R. J. and G. D. Novelli. *Arch. Biochem. Biophys.* 94: 48 (1961).

PART
10
PLANT HORMONES: BIOASSAY TECHNIQUES

INTRODUCTION

When a hormone is applied to a responsive plant system, it brings about a specific change which results, eventually, in a measurable biochemical or physiological effect. Two distinct aspects of the measured effect are involved: the specific change in metabolism, and the series of steps which lead to the physiological effect. Usually the molecular interaction of the hormone at its site of action is referred to as the "mechanism" of action. The sequence of reactions leading to the physiological effect is referred to as the "mode of action." Therefore, by definition, every hormone has its own distinctive mechanism(s) of action even though the manifestation of the hormonal mechanism may depend on prior action by other factor(s). Thus it is possible that the hormonal mechanism in one plant system may lead to a series of physiological responses that may be completely different in a second system. This difference in mode of action can be brought about because the second system has more (or less) of another hormone, has other biochemically rate-limiting components, or structural and cytological differences in tissues, etc. Therefore various bioassay techniques are more sensitive than others for a specific tissue system, and it is often advisable that several systems be used to measure the quantity of a particular hormone. In this part a brief description of the various hormones will be presented along with a number of bioassay techniques. However, to introduce the student to this field of research a brief discussion of each hormone is presented.

AUXIN

The discovery by Went (1) led to many studies of the uses of auxins in plant propagation. In 1935, Zimmerman and Wilcoxon (2) discovered that α-naphthaleneacetic acid and some chlorinated phenoxyacetic acids are very effective auxins. The strong auxin activities of these chemicals, and particularly of 2,4-D (2,4-dichlorophenoxyacetic acid), opened the way for some of the greatest practical uses of auxins.

During World War II, the herbicidal properties of the phenoxyacetic acid derivatives were carefully studied, under cover of military security,

The bioassay techniques described in Part 10 were partially devised by Dr. Brian Loveys.

FIGURE 58. Gibbane skeleton.

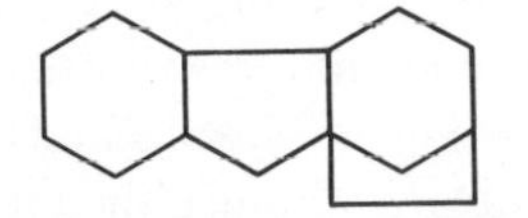

both in England and in the United States. The results of these studies were published in scientific journals in 1945 and 1947 by four research teams. The herbicidal uses of growth regulators are by far the most widespread today. It is estimated that during 1970 as many as 50,000 tons of 2,4-D were manufactured annually in the United States and perhaps 80,000,000 acres of land were treated with this herbicide alone.

Silberger and Skoog [see (3a)] were the first to report that auxin (indoleacetic acid, IAA) markedly affects RNA and DNA contents in plants. Auxin enhanced the contents of nucleic acids in tobacco pith tissue cultures on a sucrose agar medium. The increase in nucleic acid content occurred prior to the auxin-induced growth of the tissue. Concentrations of IAA that were optimal for cell enlargement and RNA synthesis often had no effect on cell division. However, lower concentrations of auxin, when compared to the optimum for cell enlargement and RNA synthesis, favored DNA synthesis and cell division.

The treatment of susceptible seedlings with herbicidal concentrations of 2,4-D produces a wide range of morphological and physiological changes. One of these is the induction of cell division in the phloem area of the soybean hypocotyl. Cell division is first noticed within 15–24 hr after 2,4-D treatment and is preceded by enlargement of the nucleus and nucleolus. Accompanying these changes is a dramatic increase in RNA content and in the activity of the enzyme involved in RNA synthesis, RNA polymerase.

At present the mechanism of auxin action is widely debated. While some workers feel that auxins directly influence nucleic acid synthesis and thereby control the physiology of the plant, many others believe the first action of auxin is at the level of the cell wall. In fact, auxin at physiological concentrations has been shown to increase growth within a few minutes. The molecular action of auxin on the cell wall is not known.

GIBBERELLINS

The term gibberellin may be defined as a compound having a gibbane skeleton (Figure 58) and biological activity in stimulating cell division or cell elongation, or both. Gibberellins also may possess other biological activities normally associated with naturally occurring substances, such as induction of nucleic acid and enzyme synthesis.

Presently 38 known gibberellins isolated from green plants or the fungus, *Gibberella fujikuori*, have been identified. Since the late 1960s three to four new gibberellins have been identified each year. Thus it seems that great progress is being made in the identification of naturally occurring gibberellins. The mechanism of action of gibberellins is still unknown. Plant physiologists are confronted with a wide array of gibberellins; sometimes several occur in the same tissue. What are these gibberellins doing within the cells? The work of Varner and Johri (4) on the induction and regulation of α-amylase appeared to offer the best approach to solving this question. At the moment, many workers in the field do not accept the idea that gibberellins regulate physiological process through the control of gene transcription (specific RNA synthesis). Good alternatives to this concept, which suggest an action of gibberellins on lipids and membranes, are being put forth now.

CYTOKININS

The term cytokinin has been accepted practically universally as a generic name for substances which promote cell division and exert other growth regulatory functions in the same manner as kinetin. The synthesis and testing of compounds for cytokinin activity began with the discovery of kinetin (6-furfurylaminopurine). Today there are probably at least a hundred known synthetic and native cytokinins.

Structural requirements for a high order of cytokinin activity generally include an adenine molecule with the purine ring intact and with a ^{6}N substituent of moderate size. Many properties of the molecule influence its activity; an apparent exception to the requirement of a modified purine exists, notably diphenylurea and its derivatives. Certain substances which lack a purine ring, such as 8-azakinetin, 6-benzylamino-8-azapurine, and 6-(3-methyl-2-butenylamino)-8-azapurine, are active, but each one of them is less than 10% as active as its corresponding purine derivative. Substitution of either O or S for N in the ^{6}N position of adenine in each case results in a more than 90% loss in growth-promoting activity in the tobacco bioassay. In comparisons of 2iP (2-isopentyladenine) with its analogs, 6-(3-methyl-2-butenylamino)-8-azapurine, 6-(3-methyl-2-butenylthio)purine, and 6-(3-methyl-2-butenylthio)-8-azapurine, the relative activities of the compounds in the tobacco bioassay are of the order of 1, 0.01, 0.01, and 0.0001, respectively. In comparison the relative activities of

6-benzylaminopurine and kinetin are 2 and 1.4, respectively. Benzimidazole and pyrimidine derivatives are active in the same manner as ⁶N-substituted adenines in some tests.

Cytokinins play a role in practically all phases of plant development, from cell division to the formation of flowers and fruits. They affect metabolism, including the activities of enzymes and the biosynthesis of growth factors. They also influence the appearance of organelles and the flow of assimilates and nutrients through the plant. Cytokinins defer senescence and may protect the plant against adverse environments, such as water stress, etc. These many diverse effects presumably stem from some primary anabolic function of the cytokinins that remains to be elucidated.

Cytokinins have been found in transfer RNA from a large number of organisms including bacteria, yeast, plants, and animals. In sequence analyses the native cytokinin, isopentenyladenine, always occurs adjacent (3′ side) to the anticodon. Furthermore, the location of the isopentenyl group is required for the tRNA to function efficiently in protein synthesis. Results involving bacteriophage infection of bacteria indicate that cytokinin moieties in tRNA may have an important regulatory function in protein synthesis. In soybean seedlings, the application of a cytokinin, 6-benzyladenine, results in increased levels of two species of leucyl-tRNAs. Although the relationship of cytokinins to tRNA may not be the only mechanism or even be the most direct action of cytokinins on cellular activity, the alteration of specific tRNAs could control protein biosynthesis. Such a regulatory mechanism would be manifested in the control of growth and morphogenesis as the ultimate expression of the hormone.

ETHYLENE

Ethylene is the simplest hormone which regulates plant growth. It is a natural constituent of plant metabolism and affects a wide array of physiological processes. For many years, however, ethylene physiology was concerned primarily with fruit ripening. With the advent of gas chromatography, many experimenters have begun research on the biochemical and physiological action of ethylene. The production of ethylene by germinating seeds and seedlings suggests that the hormone is involved in the normal regulation of growth and development. Evidence for such a proposal includes the fact that low concentrations of applied ethylene block photoinduced apical bud expansion and hook opening in etiolated pea seedlings. It was shown that apical tissues of the etiolated seedlings are the major site of ethylene production. Light has been found to decrease the tendency of pea stem segments to produce ethylene in response to high concentrations of auxin. On the other hand, red light appears to stimulate ethylene production in dormant seed. Furthermore, applications of ethylene will overcome the far-red-induced dormancy. Therefore the production of ethylene in some plant tissues is regulated by the phytochrome system (red and far-red light) and may be related to a number of light-induced phenomena.

Another important role of endogenous ethylene in etiolated seedlings is the regulation of radial expansion of the pea epicotyl in a region below the apical hook. The exposure of the plant to red light or to CO_2 inhibits an increase in the diameter of the epicotyl.

Growth, flowering, abscission, and fruit ripening—all are affected by ethylene. It is currently popular to speculate that the mode of action of ethylene involves a mechanism that regulates some aspect of transcription of DNA and translation of RNA. Studies of ethylene on abscission and growth indicate sizable changes in RNA and protein contents. There is indeed evidence of changes in activities of peroxidase, catalase, and other hydrolyases. Furthermore, a careful study showed that exposure of soybean plants to ethylene significantly alters the RNA polymerase activity associated with chromatin. As judged by the nearest neighbor analysis, the RNA produced by chromatin from ethylene-treated plants has a different base composition than that of the control.

Even though ethylene at very low concentrations affects a wide array of physiological and biochemical processes in plants, much more work is required to elucidate its mechanism of action.

ABSCISIC ACID

Abscisic acid (ABA) is a plant hormone which now ranks in importance with the auxins, gibberellins, cytokinins, and ethylene. Interest in the physiology and chemistry of ABA has grown greatly since the structure was established in 1965. During the 1950s and early 1960s a number of laboratories were engaged in research in growth-inhibiting substances. ABA was first isolated from cotton plants and was named abscisin II by a team from Addicott's laboratory (5). The structure is illustrated in Figure 59. During the same year Wareing's research team (6) isolated an active substance from *Acer* leaves that was named dormin. Abscisin II and dormin are the chemical substance which is now called abscisic acid (ABA).

FIGURE 59. *s*-Abscisic acid.

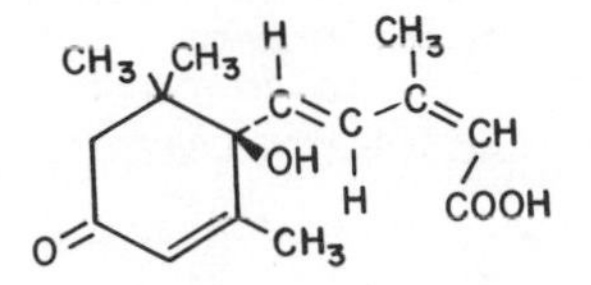

Plant tissue of all ages appears to synthesize and inactivate ABA. The number and variety of plant responses to ABA is very large. Generally, the physiological processes are related to senescence or abscission and growth retardation or inhibition. ABA appears to act as an abscission-accelerating hormone in many fruits and leaves. Furthermore, it also tends to induce dormancy in some woody plants. ABA has been shown to move from the leaves to the apical bud to bring about a dormant condition. In potato, levels of inhibitors, including ABA, decrease during the quiescent period. In addition, ABA in extremely low concentrations prolongs dormancy of excised potato buds.

Another physiological role of ABA deals with the opening and closing of stomates of leaves. Recent information shows that ABA applied through the petiole almost immediately results in stomatal closure. This mechanism of action is also not understood.

ABA inhibits germination of a number of species of seeds. It appears that a hormonal relationship dependent upon ABA concentrations exists in dormant seeds. At present the mechanism of action of ABA is not clearly understood. However, some available evidence indicates that ABA affects transcription as shown by reduced activity of chromatin-associated RNA polymerase. In other cases, the mechanism of action of ABA appears to be manifested in the translation of long-lived mRNA. The manner in which these hormones control translation is not known.

EXTRACTION AND PURIFICATION OF PLANT GROWTH SUBSTANCES. GENERAL NOTES

The success of these experiments depends on the correct choice of tissue for extraction and an appropriate bioassay for the growth substance. In general, it is best to use tissue known to be rich in the hormone of interest. Often the student may go through a long extraction procedure and end up with no biological activity. The purification steps in these experiments have been kept as simple as possible, but some of the inhibitory substances present in a crude plant extract must be removed so that the bioassay system can respond to the minute amount of growth substance present in the extract. Data given in Table 15 demonstrate the necessity of purifying the extract.

TABLE 15. DEPENDENCE OF APPARENT BIOLOGICAL ACTIVITY OF AN EXTRACT ON ITS DEGREE OF PURITY[a]

Purification step	*Specific activity* *μg GA_3 equivalents/mg extract*
Crude acidic ethyl acetate fraction of methanol extract of 48 kg of bean plants	0
Countercurrent distribution I	1.00
Countercurrent distribution II	11.35
Sephadex G-10 chromatography	274.50
Silicic acid partition chromatography	1525.00

[a] Adapted from Crozier *et al.*, *J. Exp. Bot. 20*, 786–796, 1969.

The data also show that it would be impossible to reach an accurate estimate of the actual quantity of hormone in a tissue using bioassay of a partially purified extract. An accurate estimate can be obtained only after identification of the growth substance followed by estimation by some physical means, for example, gas chromatography, spectropolarimetry. A simple method for the determination of ABA in plant tissues has been devised by Lenton *et al.* (7), using gas chromatography in conjunction with an internal standard.

It is always best to assume that the growth substance or substances in an extract are both heat- and light-labile. Extracts should always be kept as cold as possible, and undue exposure to strong light must be avoided. Extracts must be processed as quickly as possible and stored in a refrigerator.

Bottles of growth regulators should be labeled and stored in a refrigerator. Standard solutions may be stored in the cold but should not be kept for more than 6 weeks or so. It is usually best to make fresh solutions for each experiment. Most of the available growth regulators are easy to get into solution. GA_3 may be purchased as the potassium salt, in which case it will be immediately soluble in water. Raising the pH with KOH or $NaHCO_3$ will help in the solubilization of GA_3, kinetin, and IAA. The pH should be adjusted to neutrality as soon as possible. (*Note:* Kinetin comes out of solution at physiological pH at concentrations above 60 mg/l.) Gentle heat will also help in solubilization. It is safe to autoclave kinetin, but this should be avoided with IAA and gibberellins; filter sterilization should be used instead.

Organic solvents to solubilize hormones must not be used, as they can have undesirable effects in biological test systems.

Further discussion of the techniques used in growth regulators may be found in the "General References."

REFERENCES

1. Went, F. W. and K. Akad. Wetenschap. *Amsterdam Proc. Sect. Sci.* 37: 445 (1934).
2. Zimmerman, P. W. and F. Wilcoxon. *Contr. Boyce Thompson Inst.* 7: 209 (1935).
3. Cherry, J. H. *Hort. Sci.* 5: 205 (1970).
4. Varner, J. E. and M. M. Johri. In *Biochemistry and Physiology of Plant Growth Substances*. Eds. F. Wrightman and G. Setterfield. Runge Press, Ottawa. 1967, p. 793.
5. Ohkuma, K., J. L. Lyon, F. T. Addicott and O. E. Smith. *Science* 142: 1592 (1963).
6. Eagles, C. F. and P. F. Wareing. *Nature* 199: 874 (1963).
7. Lenton, J. R., V. M. Perry, and P. F. Saunders. *Planta* 96: 271–280 (1971).

General References

Addicott, F. T. and J. L. Lyon. *Ann. Rev. Plant Physiol.* 20: 139 (1969).

Agriculture Handbook No. 336. Plant Hormones and Growth Regulating Substances. USDA publication.

Key, J. L. *Ann. Rev. Plant Physiol.* 20: 449 (1969).

Klein, K. M. and D. T. Klein. *Research Methods in Plant Science*. Natural History Press. New York.

Lang, A. *Ann. Rev. Plant Physiol.* 21: 537 (1970).

Leopold, A. C. *Plant Growth and Development*. McGraw-Hill Book Co. New York. 1964.

Mitchell, J. W. and G. A. Livingston. U.S. Dept. Agr. Agric. Handbook No. 336 (1968).

Paleg, L. G. *Ann. Rev. Plant Physiol.* 16: 291 (1965).

Pratt, H. K. and J. D. Goeschl. *Ann. Rev. Plant Physiol.* 20: 541 (1969).

Skoog, F. and D. J. Armstrong. *Ann. Rev. Plant Physiol.* 21: 215 (1970).

EXPERIMENT 27 EXTRACTION AND BIOASSAY OF AUXIN-LIKE SUBSTANCES

A. OBJECTIVE

It is frequently desirable to measure the amounts of auxin-like material from plant extracts, because the content may change as a function of differentiation or as the result of various chemical or environmental conditions. In this experiment the main objectives are to learn techniques dealing with the extraction of auxin materials from plant tissue and to learn the oat coleoptile straight growth test.

B. SUPPLIES, EQUIPMENT, AND PLANT MATERIAL

Victory oat seed
Alaska pea seed or a cabbage
Sorvall blender
Methanol
80% and 95% Ethanol (cold)
Diethyl ether
Whatman No. 3 paper
Rotary evaporator
Chromatographic paper (Whatman No. 1)
Chromatography solvent: isopropanol:water (8:2 v/v)
Chromatography tank
Petri dishes
2% Sucrose, 0.02 *M* KH_2PO_4, pH 4.8–5.0
Indoleacetic acid (IAA)
Abscisic acid (ABA)
Hand lens or dissecting microscope

C. EXPERIMENTAL PROCEDURE

1. Auxin Extraction

Germinate Alaska (normal) peas in the dark in vermiculite at 25 C for 4 days. Cut about 10 g of 5-mm stem sections, including the apex, from the seedlings. Carry out this step and as much of the following as possible in subdued light. Grind the weighed sample in a blender with 50 ml of ice-cold 80% (v/v) aqueous ethanol for about 1 min. Allow the grindate to stand on ice for 1 hr, stirring occasionally. Filter the solvent through Whatman No. 3 filter paper and reduce the volume to about 8 ml with a rotary evaporator at 25–30 C. Wash out the flask with distilled water to make the final volume about 12 ml. Adjust the pH to 2.8 with 0.1 *N* HCl. Extract three times with diethyl ether in a separatory funnel. Evaporate the ether phase to dryness in a rotary evaporator and take up in about 1 ml of methanol. Use this extract for chromatography.

Another good source of auxin is the white "heart" of a cabbage. If a blender and rotary evaporator are not available, auxin-like substances can easily be extracted from this material with diethyl ether after grinding the tissue with a mortar and pestle. After allowing the extraction of the material on ice as above, remove the ether extract by filtration, using Whatman siliconized filter paper, and allowed it to evaporate in a hood. The dried extract can then be taken up in methanol and purified further by paper partition chromatography.

Better yields of biologically active material will be obtained if the extraction and purification steps are carried out in subdued light.

2. Chromatography

Strip load the extracts on Whatman No. 1 chromatography paper. Usc a piccc 10 or 12 cm widc and about 40–50 cm long. Mark the start line in pencil 10 cm from one end. A suitable loading pipet may be made by heating the end of a Pasteur disposable pipet and drawing it out to a fine point. Place the chromatogram in thc tank and allow it to equilibrate for 1 hr before adding solvent to the trough. Use isopropanol: water (8:2 v/v) as solvent. Allow the solvent front to move 25–30 cm past the origin, remove the chromatogram, mark the solvent front, and allow it to dry in a hood for 1–2 hr. Run a chromatogram with pure IAA at the same time as the other chromatograms. The spots should be visible as quenching zones when viewed in ultraviolet light. Compare these quenching zones for IAA with those on your chromatograms. Do not expose the chromatogram to U.V. light longer than necessary.

3. Coleoptile Test

A general reference for the test is given by Leopold (1). Coleoptile sections will be grown in small petri dishes on the chromatography paper with a weak nutrient solution to facilitate growth responses. Victory oat seed is soaked in running water for 1 hr to remove germination inhibitors from the hulls and then placed in vermiculite. Germinate the seeds in the dark at 25 C, giving 20–30 min of red light per day to inhibit mesocotyl elongation. When the coleoptiles are between 25 and 35 mm long, carefully select then and cut sections 5 mm long, discarding the apical 2 mm.

Cut the chromatogram(s) into 10 equal zones and place each zone in a 5-cm petri dish. Also include a control strip cut from below the origin and three similar strips to which you can add sufficient IAA in methanol to make the final concentration 10^{-5}, 10^{-6}, and 10^{-7} M IAA. Allow the methanol to evaporate and add 3 ml of a 2% sucrose (w/v) solution containing 0.02 M KH_2PO_4, pH 4.8–5.0, to each dish.

Add 10 coleoptile sections to each dish and incubate for 12–16 hr at 25 C in the dark on a gently oscillating shaker table. Then read the length as accurately as possible.

4. Bioassay of Inhibitors

The coleoptile straight growth test can also be used for the assay of plant growth inhibitors. Elute the chromatogram segments in 3 ml of water in 5-mm petri dishes for 1–2 hr. Add the coleoptile sections to each dish and incubate as in the preceding test for 1 hr, then add 1 ml of incubation medium so that the final concentration of the components is as follows:

Sucrose	2% (w/v)
KH_2PO_4	0.02 *M*
IAA	5×10^{-6} *M*

Measure coleoptile length after 12–16 hr of incubation at 25 C in the dark on a shaker. A standard series of abscisic acid (ABA) concentrations (10^{-7} to 10^{-4} *M*) may be included if available.

Note: Etiolated tissue usually contains very low levels of growth inhibitors. Therefore, choose light-grown tissue for the extraction of inhibitors. Almost any vegetative green plant material will show the so-called "β-inhibitor" zone on chromatograms (R_F 0.5–0.8 of chromatograms developed in isopropanol:ammonia:water, 10:1:1 v/v). Use 5–8 g fresh weight of tissue to demonstrate this inhibitor with the coleoptile test. The bulk of the β-inhibitor zone is thought to be due to ABA. The ABA content of tissue may be increased considerably by allowing the tissue to wilt before extraction (10–15% loss of fresh weight for 4 hr). An excellent source of ABA is the fruit of the rose. The pseudocarp of *Rosa arvensis* may contain up to 4.1 mg/kg fresh weight (2).

D. TREATMENT OF RESULTS

Present your results as a short report including a discussion of auxin promoters and inhibitors.

REFERENCES

1. Leopold, A. C. R. *Auxins and Plant Growth.* Univ. Calif. Press. 1955, p. 34.
2. Milborrow, B. V. In *Biochemistry and Physiology of Plant Growth Substances.* Eds. F. Wightman and G. Setterfield. Runge Press, Ottawa. 1967, pp. 1531–1545.

General Reference

Nitsch, J. P. and C. Nitsch. *Plant Physiol.* 31: 94 (1956).

EXPERIMENT 28 EXTRACTION AND BIOASSAY OF GIBBERELLIN-LIKE SUBSTANCES

A. OBJECTIVE

Gibberellins affect an array of physiological processes ranging from stem elongation to flowering and fruit setting. The objective of this experiment is to learn several techniques for the estimation of gibberellic acid (GA) content.

B. SUPPLIES AND MATERIALS

Barley seed
Great Lakes or Arctic lettuce seed
Alaska pea seed
Rumex plants
Blender
80% (v/v) Methanol (cold)
Ethyl acetate
Rotary evaporator
Chromatography equipment (see Experiment 27)
Petri dishes
GA_3 solutions (10^{-8}, 10^{-7}, 10^{-6}, and 10^{-5} *M*)
Spectrophotometer
1% Methylene blue
1% Sodium hypochlorite
Starch reagent: 150 mg of potato starch (not solubilized), 600 mg of KH_2PO_4, 29 mg of $CaCl_2$, and water to make 100 ml volume; boil for 1 min; cool and filter
Iodine stock solution: 6 gm of KI, 600 mg of iodine; water to make 100 ml volume stock solution; add 1 ml to 100 ml 0.05 *N* HCl for final iodine reagent
Chloramphenicol 10 μg/ml

C. EXPERIMENTAL PROCEDURE*

1. Extraction of GA-Like Substances

The content of GA-like substances in a plant organ varies considerably with its stage of development. Good yields can normally be obtained from young meristematic tissue (for example, apices of light-grown bean or sunflower plants) or from developing bean fruits.

Homogenize 10–12 g of tissue in about 100 ml of cold 80% methanol. Allow the homogenate to stand on ice for 1 hr while stirring occasionally. Filter the extract and reduce it to the aqueous phase on a rotary evaporator (25–30 C). Freeze the aqueous extract. Thaw and then centrifuge at 20,000 × *g* for 1 hr to remove much of the pigment. Adjust the pH of the supernatant to 2.5 with 1 *N* HCl and extract three

* The gibberellin bioassays mentioned here are by no means exhaustive. For a comparison of several bioassay systems and the potency of different giberellins in each, see Crozier *et al.* (1).

times with ethyl acetate. Add anhydrous sodium sulfate (10 g/100 ml) to the extract to dry the ethyl acetate. Reduce the filtered extract to dryness with a rotary evaporator and redissolve the extract in a known volume of methanol (1–2 ml).

(a) Chromatography. The extract can be chromatographed in the same way as the auxin extract (Experiment 27). Use isopropanol: ammonia: water (10:1:1 v/v) as a solvent. Pure GA_3 (10 μg) should be run on a separate chromatogram. The location of GA can be visualized by spraying the paper with ethanol: sulfuric acid (95:5 v/v) and heating for a few minutes at 100 C. The GA_3 will appear as a blue fluorescent spot under U.V. light before the paper dries.

2. Rumex Bioassay

Senescence Deferral Test. (a) Cut the oldest leaves from *Rumex* plants, but avoid leaves that are already irregular in color. Let them stand overnight at 25 C in the darkroom with the petioles dipped in water. (b) Cut leaf discs with a No. 5 cork borer. Then randomize the discs and place five on a 2.5-cm disc moistened with 0.5 ml of solution to be tested, contained in a small petri dish. Let stand in the darkroom for 5 days, the time required for chlorophyll to disappear from the control leaf discs. Freeze two lots of five discs each in order to determine the initial level of chlorophyll (see Experiment 9 for instructions on chlorophyll determination) at the termination of the other assays.

3. The Test Solutions

These solutions shall consist of: ten zones from a paper chromatogram of an extract from 5 g of tissue and various concentrations of GA (10^{-8}, 10^{-7}, 10^{-6}, and 10^{-5} *M*) and water.

4. Lettuce Hypocotyl Test

(a) Germinate Great Lakes or Alaska lettuce seeds on wet filter paper (5 ml of water in a large petri dish with two paper discs) for 24 hr on a lighted lab bench. (b) When the white radicle is clearly visible, transfer uniformly germinated seedlings (12 per dish) to a filter paper disc or paper chromatogram section moistened with 2 ml of water in a small petri dish. Gibberellin standards of 10^{-7}, 10^{-6}, and 10^{-5} *M* GA will be employed. Place the several petri dishes in a large plastic box containing wet paper towels and having a tight lid. Place the box under a light source (1000–2000 ft-c) for 3 days. Control hypocotyls should grow to only about 2 mm, while those treated with GA at 10^{-5} *M* should grow to about 15 mm. (c) Measurement to the nearest half millimeter can be made with a simple millimeter rule. Placing two drops of 1% methylene blue on the paper before reading will facilitate easy detection of the junction of root and stem, as the root portion will turn blue. Measure from the junction of stem and root to the first node, which is seen as a slight swelling just below the cotyledonary leaves. Record your data as millimeters of hypocotyl length.

5. Barley Seed Assay

α-Amylase Secretion. (Use 1–5 g of fresh weight equivalent here.) (a) Cut barley seeds in half, discarding embryo end. (b) Sterilize the half-seeds in 1% Na hypochlorite for about 30 min. Then rinse three times in sterile (autoclaved) water. (c) Place the half-seeds in sterile sand-filled petri dishes (4-in.) and add 20 ml of sterile water to each dish. (d) After 3 days, transfer (sterilely) five seed pieces into each of several 25-ml Ehrlenmeyer flasks containing 10^{-3} *M* sodium acetate, pH 4.8, 10^{-2} *M* $CaCl_2$, 10 μg/ml of chloramphenicol, and the GA extract or GA standard—all in 2 ml. (Several concentrations of the extract and GA_3 (10^{-8} to 10^{-5} *M*) should be tested.) (e) Incubate the samples of half-seeds for 24 hr in a shaker at 25 C. (f) Gibberellins induce the synthesis of α-amylase, which is then secreted into the incubation medium. The activity of α-amylase in the incubation medium will be assayed as follows: Add to test tubes 0.2 ml of the incubation solution* from one of the test samples [see (e) above], 1.0 ml of starch reagent, and 0.8 ml of water. Incubate for 10 min at 30 C. Stop the reaction by adding 1 ml of iodine reagent followed by 5 ml water. On a spectrophotometer, read the blue color at 620 mμ.

D. TREATMENT OF RESULTS

Present your results as a comparison of the sensitivity of the two bioassay techniques. On the basis of the amount of growth responses from GA_3, calculate the amount of gibberellin extracted. Report the data in the form of a short paper for *Plant Physiology*.

REFERENCE

1. Crozier, A., C. C. Kuo, R. C. Durley, and R. P. Pharis. *Can. J. Bot.* 48: 867 (1970).

General References

Frankland, B. and P. F. Wareing. *Nature* 185: 255 (1960).
Jones, R. and J. E. Varner. *Planta* 72: 155 (1967).
Phinney, B. O. *Proc. Natl. Acad. Sci. U.S.* 42: 185 (1955).
Whyte, P. and L. C. Luckwill. *Nature* 210: 1360 (1966).

* Volume of incubation solution and incubation time can be varied.

EXPERIMENT
29 BIOASSAYS FOR CYTOKININS

A. OBJECTIVE

Since the isolation and characterization of kinetin in 1955 by Miller (1) a fast and simple bioassay technique has been required. The experiment is planned to provide the student with information on several cytokinin bioassay techniques.

B. SUPPLIES AND MATERIALS

Xanthium seeds
Xanthium plants
Razor blade
10^{-2} *M* phosphate buffer, pH 6.0
Petri dishes
Plastic box or large bell jar
N-6-Benzyladenine
2.5-cm Paper discs
Zeatin
80% Ethanol
Spectrophotometer

C. EXPERIMENTAL PROCEDURE

1. Xanthium Cotyledon Bioassay

Cytokinin-Stimulated Cotyledon Growth.

(a) Remove seeds from the burrs of *Xanthium pensylvanicum* in either of two ways: Either freeze the burrs in liquid nitrogen and break them open with a sharp blow, or soak the burrs in water for about 2 hr and cut them open with a razor blade. Be sure to sort for uniform size and color of seeds.

Each seed is reduced to about 1 cm in length by cutting off the two ends. The center piece is then cut into three sections about 3 mm each, and a longitudinal cut results in six pieces of cotyledon. These pieces are placed on wet filter paper for 5–7 hr to complete water imbibition, after which the parts of the two cotyledons separate and yield twelve pieces from each seed.

(b) Randomize these pieces and place ten in each dish for assay. Fresh weights are taken for the group of ten pieces after gentle blotting on fresh filter paper. Place the weighed pieces into small petri dishes containing three filter paper discs (4.5 cm) moistened with 1.5 ml of 10^{-2} *M* phosphate buffer, pH 6. Chromatogram pieces can be used instead of the filter paper discs. For a cytokinin standard, use 6-benzyladenine and zeatin ranging in concentration from 10^{-7} to 10^{-4} *M* in the buffer medium.

(c) Hold the petri dishes in a larger plastic box with a lid and lined with wet filter papers to prevent drying out. Place inside an oven with the temperature kept at 30 C for 4 days. Controls should increase in fresh weight (blotting gently as before) by about 150%; benzyladenine at 10^{-5} *M* should increase the fresh weight by 500%. The sensitivity of this assay runs down to 10^{-8} *M* 6-benzyladenine.

2. Xanthium Leaf Disc Bioassay: Cytokinin Deferral of Senescence

(a) Cocklebur plants (*Xanthium pensylvanicum* Wall.) are planted and grown in the greenhouse under a 20-hr photoperiod. Within about 6 weeks the plants will have about ten leaves. Remove the fully expanded leaves, usually the fifth, sixth, and seventh (counting from the apex), and place the petioles in flasks containing water. Store the leaves in Plexiglas boxes or large bell jars under low light intensity (50 ft-c) at about 20 C for 3 days. During this period, metabolic processes leading to senescence are initiated and a fall in total chlorophyll content begins. The final selection of suitable leaves is made on the basis of a uniform pale green color.

(b) Using a No. 5 cork borer, remove uniform leaf discs from the "aged" *Xanthium* leaves. Care should be taken not to include heavy veins in the leaf discs. Place four leaf discs on a 2.5-cm paper disc so that the abaxial surface is in contact with the paper, and moisten with 0.5 ml of a test solution (cytokinin solution or plant extracts). When cytokinins are chromatographed, pieces of the paper chromatogram of suitable sizes can be used. In this experiment, place several of the paper discs, each containing four *Xanthium* leaf discs, in petri dishes. Test the effect of two cytokinins, 6-benzyladenine and zeatin, ranging in concentrations from 10^{-7} to 10^{-4} *M* as compared to water controls (run in duplicate).

Place the petri dishes in a container with moist paper toweling or in some suitable humidity chamber. Incubate the sample in this moist environment at 24 C in darkness for 48 hr.

(c) At the initiation of the incubation of the leaf discs, extract the chlorophyll of a representative group of four discs by dropping them into tubes containing 5–6 ml of boiling 80% ethanol. Boil gently until the green color is extracted from the discs, then cool and bring the volume to 10 ml with 80% ethanol. Read the absorbance at 665 mμ.

At the termination of the 48-hr incubation period, remove the leaf discs from the paper discs and extract the chlorophyll as indicated above. The cytokinin will defer senescence of the leaf discs by retarding chlorophyll loss.

3. Other Cytokinin Bioassay Techniques

The tobacco pith callus bioassay technique developed by Skoog and Miller (2) has been extremely useful in the survey of cytokinins. The Wisconsin group has used this technique almost exclusively. However, the tobacco callus requires several weeks for the test to be completed. Miller (3) developed another bioassay technique which utilizes the soybean callus. The soybean test is easier than the tobacco test, but it also requires several weeks for assay. Recently Letham (4) has described a "rapid" radish cotyledon bioassay technique which requires 3 days for completion. This bioassay will detect kinetin at the low level of 10 μg/liter.

D. TREATMENT OF RESULTS

Compare the sensitivity of the *Xanthium* cotyledon and *Xanthium* leaf disc bioassay techniques on the basis of your results. A brief report containing relevant tables and graphs should also provide information on the possible improvement of the techniques.

REFERENCES

1. Miller, C. O. *Proc. Natl. Acad. Sci. U.S.* 54: 170 (1965).
2. Skoog, F. and C. O. Miller. *Symp. Soc. Exp. Biol. Med.* 11: 118 (1957).
3. Miller, C. O. In *Modern Methods of Plant Analysis*. Vol. 6. Eds. K. Paech and M. V. Tracey. Springer-Verlag, Berlin. 1963, p. 194.
4. Letham, D. S. In *Biochemistry and Physiology of Plant Growth Substances*. Eds. F. Wrightman and G. Setterfield: Runge Press, Ottawa. 1967, p. 19.

General References

Esahi, Y. and A. C. Leopold. *Plant Physiol.* 44: 618 (1969).
Osborne, D. J. and D. R. McCalla. *Plant Physiol.* 36: 219 (1961).

EXPERIMENT 30 ETHYLENE BIOASSAY BASED ON PEA STEM SWELLING

A. OBJECTIVE

Ethylene affects a number of physiological processes including fruit ripening, bean hypocotyl hook opening, and seed germination. Because of the precision of gas chromatographic determination of ethylene, there has been no widely accepted bioassay technique providing an accurate estimation of ethylene. However, a very useful ethylene bioassay technique has been developed in the Horticulture Department at Purdue (1). The objective of this experiment is to learn the pea stem swelling technique as related to the measurement of ethylene.

B. SUPPLIES AND MATERIAL

Alaska pea seeds
Vermiculite
Ehrlenmeyer flasks
India ink

C. EXPERIMENTAL PROCEDURE

1. Plant Alaska pea seeds in vermiculite and allow them to germinate in the dark (25 C) for 4 days. At this time the stems should have a height of about 2 or 3 cm.

2. Remove the roots by cutting just below the cotyledons. Mark the epicotyl with an India ink pen at a position 1 cm below the apex.

3. Place ten of the derooted cuttings in a 50-ml Ehrlenmeyer flask with 5 ml of water or the solution being tested, and cap with a rubber serum cap. Using a hypodermic syringe, add volumes of ethylene gas from a stock sample to provide concentrations of 0, 0.1, 1, 10, and 100 ppm of ethylene (v/v) in the flasks. In addition, compare the effects of these concentrations of ethylene to the effect of 10^{-3} *M* methionine. Gas chromatography may be used to determine accurately the ethylene concentration of the stock sample.

4. Swelling is measured by cutting the epicotyls off at the India ink marker and measuring individual lengths and fresh weight (blotted) of the group of apical pieces.

Incubate the samples for 24 hr. The initial weight of the 1-cm apex is found by cutting ten epicotyls at the India ink mark and weighing. The presence of the cotyledons is essential for good swelling response to ethylene. The results are expressed as a ratio of weight to length. Ratios for a control cutting (no ethylene) should be about 3.0; 1 ppm of ethylene should increase this ratio to about 4.0.

D. PRESENTATION OF RESULTS

Write a brief report of your data, including a graphic plot of stem-weight-to-length ratio versus ethylene concentration. Also include a table of all measurements obtained. Comment on the usefulness of this bioassay technique in relation to others of which you are aware.

GENERAL REFERENCE

Warner, H. Ph.D. thesis, Purdue University. 1970, p. 53.

PART 11

ANALYSIS OF PLANT CONSTITUENTS BY GAS CHROMATOGRAPHY

INTRODUCTION

Gas chromatography is a subclass of the general methodology of chromatography where separation of mixtures is obtained by multiple or repeated distribution between two phases. In this case the two phases consist of a gas (carrier gas or mobile phase) and a solid (stationary phase), or a gas and a liquid. Gas-solid chromatography (GSC) is used mainly for separation of gas mixtures (ethylene, CO_2, CO, N_2, O_2), while gas-liquid chromatography (GLC) finds a wide application for the separation of compounds ranging in boiling points from −50 C to 400 C. A wide variety of materials may be used as the liquid phase: higher hydrocarbons, esters, polyesters, polyethers, silicones, etc. Selection of the proper liquid phase is very important and depends on the nature of the separation problem at hand. The substance used must be chemically stable at the operating temperature and must have a low volatility to minimize losses into the carrier gas.

In GLC the liquid is deposited as a thin film on small particles of a solid support. The support is composed of specially treated inert material, such as diatomaceous earth or crushed firebrick, selected to provide a large surface area and to give a minimum of interaction with solute molecules (sample). This coated material is put into tubes of glass or metal (stainless steel or copper) and constitutes the gas chromatographic column, the heart of the gas chromatograph (Figure 60).

A carefully regulated flow of an inert gas (N_2 or He) is constantly passing through the column. The sample is introduced through an injection port, usually by means of a microsyringe, into a heated area where it is rapidly volatilized and swept by the carrier gas into the column. By interaction of the solute with the liquid phase and because of differences in the solubility or affinity of the sample components for the liquid phase, separation takes place (Figure 61).

As the separated components emerge from the column, their presence in the carrier gas is sensed by the *detector*. Many types of detectors are used in gas chromatography. One of the most common, and the one

Analysis of plant constituents by gas chromatography was devised by Prof. Johan Hoff.

FIGURE 60. Diagram of a general gas chromatograph.

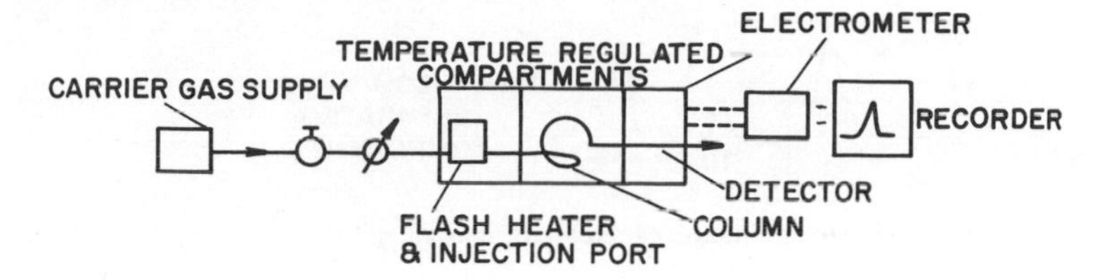

FIGURE 61. Separation of two compounds by gas chromatography.

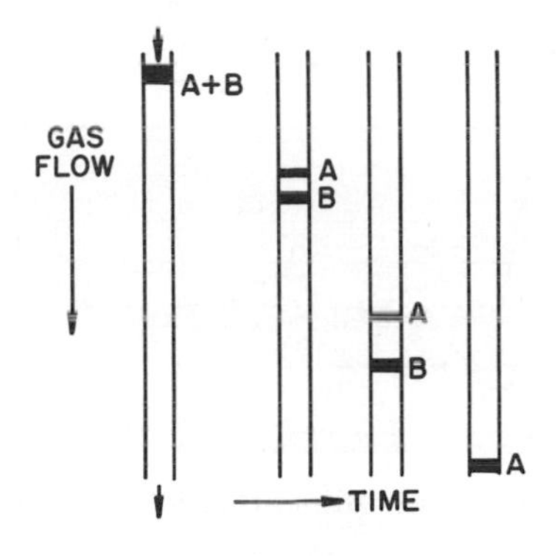

to be used in these experiments, is the flame ionization detector. Here hydrogen gas is introduced into the carrier gas stream at the outlet of the column. The gas mixture emerges through a fine tip into the detector chamber proper, where the hydrogen feeds a burning flame maintained by air that is also introduced into the chamber. The flame tip and a *collector ring* which surrounds the flame constitute two electrodes between which an electric potential is imposed. In the presence of ions in the flame an electric current will pass through the system. The amount of ions is enormously increased when a compound emerges from the column and is combusted in the flame. The resulting increase in current is amplified by the electrometer, and the amplified signal is fed to the recorder. The gas chromatogram is a recording of the signal produced by the detector, as indicated in Figure 62.

The detector produces time-dependent deviations from the base line when a signal is transmitted. In a good detector the signal is proportional to the concentration of the eluted components over a wide range of concentrations. Such a detector is said to have considerable linear dynamic range. This is true of most hydrogen flame detectors, and it accounts in part for the popularity of this type of detector. Another valuable characteristic of the flame ionization detector is its insensitivity to water. It is therefore possible to inject samples of dilute aqueous

FIGURE 62. Recording of materials separated by gas chromatography.

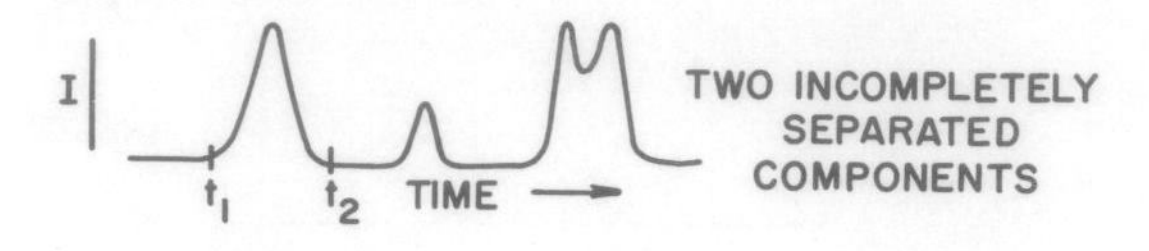

FIGURE 63. Determination of retention time, $t_{R_{1/2}}$.

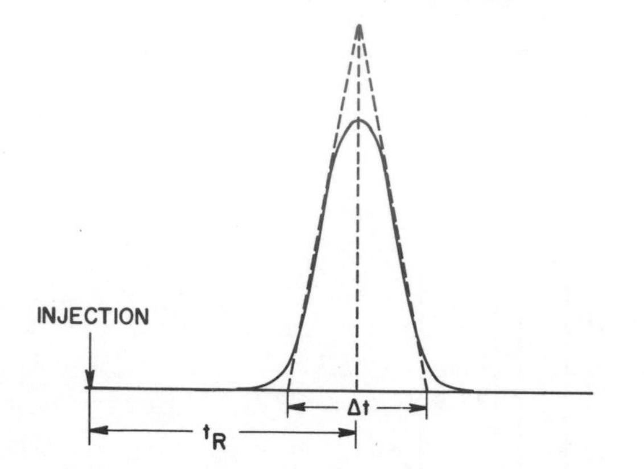

solutions and obtain a chromatogram undisturbed by the relatively large amount of water in the sample.

If the intensity of the signal (magnitude of the recorder deflection) is proportional to the concentration,

$$\text{Intensity } (I) \sim \text{concentration } (C),$$

then

$$\text{Amount} = \int_{t_1}^{t_2} C\,dt \sim \int_{t_1}^{t_2} I\,dt = \text{area } (A) \text{ under curve}$$

That is,

$$\text{Amount} \sim A$$

The chromatogram therefore provides quantitative information. The area under the peak is conveniently estimated by an automatic integrator, which often is provided with the recorder. The peak height is sometimes, particularly for narrow symmetrical peaks, proportional to the amount.

The chromatogram also provides qualitative information. If the separation is adequate, it tells how many components there are in the mixture. Under a given set of circumstances the various components emerge at reproducible and characteristic times from the column. The time from injection point to the center of the peak, the retention time or t_R, is analogous to R_F in paper chromatography and to the elution volume in column chromatography and is useful in giving clues to the identity of unknown compounds (Figure 63).

The separation efficiency of a column often determines whether a particular experiment is a success or a failure. It is influenced by several factors such as particle size of the inert support, the type and amount of liquid phase used, the parameters chosen for the experiment (i.e., temperature, length and cross section of the column), and the flow rate of the carrier gas. It is therefore important to evaluate the ability of a column to separate compounds in order to have assurance of adequate separation. The number of theoretical plates, N, is a convenient way of designating the separating efficiency of a column:

$$N = 16\left(\frac{t_R}{\Delta t}\right)^2$$

Since separation is affected by the fact that the compounds are present in the gas phase at least part of the time, a prime requisite for the application of GLC is that the compounds to be separated possess a certain volatility, and that the volatility be sufficient to bring about a reasonable forward movement within the column.

Many compounds of interest in the life sciences are not sufficiently volatile or become too unstable at high temperatures to make GLC a practical tool for their separation. Such compounds (amino acids, fatty acids, carbohydrates, phenols, steroids) owe their low volatility to the polarity of their active groups, tending to form intermolecular hydrogen bonds. The polarity of such groups can be minimized by forming certain derivatives of the compounds, i.e., covering the active groups by reaction with groups of lesser polarity. The resultant derivatives, having a reduced tendency to form intermolecular hydrogen bonds, exhibit increased volatility and can be separated by GLC without suffering thermal decomposition. Thus, for free amino acids the *N*-trifluoroacetyl butyl esters and for free fatty acids the corresponding methyl esters are formed before separation by GLC. Carbohydrates are first reduced to the corresponding alditols and then acetylated to polyacetates.

EXPERIMENT

31 DETERMINATION OF THE FATTY ACID COMPOSITION OF LEAF LIPIDS

The lipids of green leaves are found primarily in membranes of various cell structures such as the plasmalemma, organelle membranes (nucleus, mitochondria, chloroplasts), and endoplasmic reticulum. Therefore the quantity and composition of the lipids reflect to some extent the metabolic activity of the tissues and can conveniently be used to assess effects on the cell metabolism of light intensity, temperature, or other environmental factors. This experiment introduces the student to commonly used techniques of lipid extraction and determination of the fatty acid composition by GLC. These techniques are applicable to a wide variety of materials of biological origin.

A. EQUIPMENT AND SUPPLIES

A gas chromatograph equipped with dual columns and flame ionization detectors, temperature-programmed oven, and direct on-column injection system.

Columns. Use glass columns if possible, 1.5 m × 4 mm i.d., with the substrate, diethylene glycol succinate (DEGS), deposited on Chromosorb W, acid-washed, 80–100 mesh. Columns may be prepared and properly cured according to methods described by Burchfield and Storrs (1) or purchased ready-made from gas chromatography supply houses.

Lipid extractant (solution A). Prepare 1 liter of a 2:1 (v/v) mixture of chloroform and methanol containing 0.2% (butylated hydroxytoluene, an antioxidant).

5% Anhydrous HCl in methanol (solution B). Operating in a ventilated hood, bubble a slow stream of anhydrous HCl through 1 liter of absolute methanol in a glass-stoppered amber bottle (with the stopper removed). At intervals remove 5.0-ml aliquots and titrate with 0.1 *N* NaOH to determine the concentration of HCl. Continue adding HCl until the proper concentration has been reached, taking care that the stream is vigorous enough to prevent back flushing of solvent, yet slow enough to make efficient use of the gas.

Standard solution (solution C). Prepare a solution of the methyl esters of the authentic fatty acids listed in Table 16, containing 1 mg/ml of each compound.

TABLE 16

	No. of Carbon Atoms	No. of Double Bonds	Code
Myristic acid	14	0	14:0
Palmitic acid	16	0	16:0
Palmitoleic acid	16	1	16:1
Stearic acid	18	0	18:0
Oleic acid	18	1	18:1
Linoleic acid	18	2	18:2
Linolenic acid	18	3	18:3
Eicosanoic acid (arachic)	20	0	20:0
Eicosenoic acid	20	1	20:1
Docosanoic acid (behenic)	22	0	22:0
Docosanoic acid	22	1	22:1
Tetracosanoic acid (lignoceric)	24	0	24:0
Tetracosanoic acid	24	1	24:1

B. EXTRACTION OF CRUDE LIPIDS

Freeze preweighed 1-g portions of green leaves directly on a Dry Ice block. When fully frozen, transfer to a pre-cooled (on Dry Ice) 50-ml mortar and pestle and grind to a fine powder. Add 5 ml of solution A to the ground powder in a test tube. Swirl or shake repeatedly during 1 hr at room temperature and siphon off 2 ml of the bottom layer through a cotton-tipped pipet. Empty the pipet into a test tube of suitable size with a Teflon-lined screw cap, and evaporate the solvent at 50 C from the open tube in a stream of pure nitrogen.

C. INTERESTERIFICATION OF LIPIDS

Dissolve the crude lipids in 4 ml of solution B, add 0.5 ml of benzene, flush with nitrogen, and screw the cap on firmly. Heat in oil bath at 70 C for 2 hr. Pour the contents into a larger test tube, washing the small test tube with several small portions (total 7 ml) of petroleum ether during the transfer. Add 8 ml of water. Stopper the test tube and shake carefully to avoid formation of an emulsion. (If an emulsion is formed, it may be broken by sprinkling finely ground Na_2SO_4 over the surface.) Let the layers separate, and siphon off the top layer carefully, avoiding contamination with water. Save the extract and repeat the extraction with 7 ml of petroleum ether twice. Transfer the combined extracts to a separatory funnel and dry and neutralize over a mixture of solid sodium sulfate and sodium bicarbonate. Transfer to a test tube and reduce the volume of the solution to approximately 2 ml in a stream of pure nitrogen.

D. GAS CHROMATOGRAPHY

Inject 10 μl of solution C into the chromatograph. Operate the column oven isothermally at 150 C and 190 C in two separate runs or, if programming is available, at an initial temperature of 150 C, increasing at a rate of 2°/min to a maximum of 200 C. The flash heater and detector blocks should be maintained at 220 C in either case. The compounds will emerge in the order listed in Table 16. Inject 10 μl of the unknown after the standard solution. Record retention times of both runs and identify the peaks in the unknown by comparison with the standard. Obtain the composition of the fatty acid mixture by calculating the proportion of each peak area in the total area comprising the sum of all peak areas.

REFERENCE

1. Burchfield, H. P. and E. E. Storrs. *Biochemical Applications of Gas Chromatography*. Academic Press, New York. 1962.

EXPERIMENT

32 MONOSACCHARIDES OF THE PLANT CELL WALL

The plant cell wall is composed of cellulose, lignin, hemicellulose, pectic substances, and protein. Among these groups of compounds, the pectic substances are the least characterized and are the most complex, varying both quantitatively and qualitatively during the various physiological processes of growth and development. This experiment attempts to characterize the monosaccharide composition of material extracted from plant cell walls (primarily pectin). Extraction of the pectic substances and hydrolysis of the polysaccharides (galactans and arabaninans) are accomplished in one operation by treatment with trifluoroacetic acid. Formation of stable and relatively nonpolar derivatives then follows (1):

Reaction 1: $$\underset{\text{Monosaccharide}}{C_nH_{2n}O_n} + \underset{\text{Borohyride}}{NaBH_4} \rightarrow \underset{\text{Alditol}}{C_nH_{2n+2}O_n}$$

Reaction 2: $$C_nH_{2n+2}O_n + \underset{\text{Acetic anhydride}}{(AcO)_2} \rightarrow \underset{\text{Polyacetate}}{C_nH_{n+2}O_n(AcO)_n} + \underset{\text{Acetic acid}}{n\,AcOH}$$

The resulting mixture of polyacetates is separated and evaluated quantitatively by means of GLC.

A. EQUIPMENT AND SUPPLIES

A gas chromatograph as in Experiment 31.

Columns. Purchase custom-made columns from a gas chromatography supply house or prepare as follows: Prepare a 50-ml solution in acetone of 0.8% ethylene glycol adipate (EGA) and 1.6% XF 1150 (a silicone grease) and a solution (50 ml) of chloroform and 0.8% ethylene glycol succinate (EGS). Mix the two solutions and immediately pour the mixture over 50 g of Gas-Crom P (Applied Science Laboratories, Inc.), 100–200 mesh, in a fritted glass funnel. Remove excess solvent by gently suction and dry the coated particles in a stream of warm nitrogen gas entering from the bottom of the funnel. Pour the dry column material into copper tubing, 1.2 m × 3 mm i.d. while tapping the tubing gently. Cure the columns overnight with a slow stream of carrier gas in the column oven at 200 C while the exit ends are disconnected from the detector.

Ultrasonic bath (ultrasonic cleaner, Model 8845-2, Cole-Parmer Instrument Co.)

Waring Blendor

Fritted glass filtration assembly, 20 mm diameter

Glass fiber filter pads (No. 934-H Reeve Angel)

Oxygen torch

Nitrogen gas cylinder

Temperature-controlled water bath

Miracloth (Chicopee Mills, Inc.)

Potato tubers

Sodium borohydride

Grinding medium (sodium A): prepare 1 liter of 0.1 *M* phosphate buffer, pH 7.0, containing 0.01% Na_2SO_3

Pectin extractant (solution B): prepare 50 ml of 2 *N* trifluoroacetic acid containing 0.5 mg myoinositol/ml as internal standard

Standard solution (solution C): prepare 200 ml of 1 *N* NH_4OH containing 0.5 mg/ml of each of the following substances: rhamnose, fucose, arabinose, xylose, galactose, mannose, glucose, and myoinositol

B. EXPERIMENTAL PROCEDURE

1. Isolation of Cell Walls

Approximately 20 g of 1-mm-thick raw potato tuber slices is suspended in cold water and the free starch washed away. Grind the washed slices in 50 ml of solution A and some crushed ice until most of the cells are broken. If the temperature exceeds 5 C during the grinding, add more ice before the grinding is continued. Inspect the material microscopically using Lugol's iodine solution. When this material appears sufficiently disintegrated, transfer the slurry to a 30 × 30 cm piece of Miracloth deposited in a funnel. Hold the sides and corners of the Miracloth and form a bag. Dip this bag repeatedly in a container of distilled water, taking care to allow almost all the water to drain between dippings. Change the water in the reservoir at intervals. Continue this treatment until no more starch appears in the wash water. Unbroken cells at this point should be few and mainly consist of sclereids. If a substantial amount of parenchymatous cells is still intact, treat the cell wall slurry ultrasonically until the cells break, and repeat the washing treatment. Transfer the virtually starch-free cell wall suspension to a 20-mm filter tube equipped with a similar Miracloth disk and wash the cell walls repeatedly, first with ethanol, then with ether.

C. HYDROLYSIS OF THE CELL WALL MATERIAL (1)

Weigh 10–20 mg of the dried cell walls and transfer to a 13 × 100 mm test tube. Add 2 ml of solution B. Seal the test tube in an oxygen flame while flushing with H_2, and hydrolyze for 1 hr at 121 C. After cooling to room temperature, open the test tube and filter through preweighed glass fiber filter pads into another test tube. Save the filtrate, wash the filter three times with small amounts of ethanol, and discard the washings. Dry and weigh the filter plus residue to determine unhydrolyzed material.

D. FORMATION OF ALDITOLS

Evaporate the filtrate in a stream of filtered air in a water bath at 50 C. Add 0.5 ml of 1 *N* NH_4OH and 10 mg of $NaBH_4$ (sodium borohydride) and leave at room temperature for 1 hr. Destroy excess $NaBH_4$ by the drop by drop addition of glacial acetic acid until effervescence ceases.

E. FORMATION OF ACETATES

Evaporate the tube contents to dryness in a stream of air at 50 C, add 1 ml of methanol, and again bring the tubes to dryness. Repeat the addition of methanol and evaporation four more times. Add 1 ml of acetic anhydride, seal the tubes under N_2, and let acylation proceed at 121 C for 3 hr.

F. STANDARDS

Prepare the derivatives, starting at Section D, with 2 ml of solution C.

G. GAS CHROMATOGRAPHY

Inject 10 μl or less of the samples while programming the temperature from 120 to 180 C at 2°/min after an initial delay of 11 min. The injection block temperature should be 220 C and the detector temperature, 270 C.

H. CALCULATIONS

The sugars emerge in the sequence indicated (see solution C). Obtain peak areas by planimetry or other means (disc integrator, electronic integrator). The peak areas on the standard chromatogram represent 1 mg of each of the sugars. Therefore the quantities of the various sugars present in the final sample solutions are expressed by

$$\text{mg X} = \frac{(A_X/A_I)_{\text{sample}}}{(A_X/A_I)_{\text{standard}}}$$

where X = sugar, I = inositol, and A stands for peak areas. Then

$$\%\text{X} = \frac{\text{mg X}}{\text{mg W}} \times 100$$

$$\%\text{X}' = \frac{\text{mg X}}{\text{mg (W} - \text{UH)}} \times 100$$

where %X is of the percentage of sugar X in the cell wall, %X', the percentage of sugar X in the hydrolyzed material (primarily pectin), and W and UH represent the weight of cell walls and unhydrolyzed material, respectively.

REFERENCE

1. Albersheim, D., D. J. Nevins, P. D. English and A. Karr. *Carbohydrate Res.* 5: 340 (1967).

EXPERIMENT

33 AMINO ACID COMPOSITION OF PLANT PROTEINS

The amino acid composition of proteins is usually obtained by ion exchange separation of their acid hydrolysates. Rather expensive automatic equipment is available for this purpose. An alternative, less expensive, and quicker method has long been sought to replace the standard procedures. The present procedure (1) makes amino acid composition studies available to laboratories that possess a suitably equipped gas chromatograph. The analysis proceeds in principle as follows: Plant proteins are isolated and hydrolyzed into individual amino acids. The polar groups of the amino acids are then rendered relatively nonpolar in a sequence of two reactions to produce stable and volatile derivatives:

Reaction 1:

$$\underset{\alpha\text{-Amino acid}}{R \cdot \underset{\substack{\cdot \\ NH_2}}{CH} \cdot COOH} + \underset{n\text{-Butanol}}{HO \cdot (CH_2)_3 \cdot CH_3} \xrightarrow{HCl} \underset{\alpha\text{-Amino acid-}n\text{-butyl ester}}{R \cdot \underset{\substack{\cdot \\ NH_2}}{CH} \cdot COO \cdot (CH_2)_3 \cdot CH_3}$$

Reaction 2:

$$R \cdot \underset{\substack{\cdot \\ NH_2}}{CH} \cdot COO(CH_2)_3CH_3 + \underset{\text{Trifluoroacetic anhydride}}{(CF_3CO)_2O} \rightarrow \underset{N\text{-trifluoroacetyl-}\alpha\text{-amino acid-}n\text{-butyl ester}}{R \cdot \underset{\substack{\cdot \\ HN \cdot (CF_3CO)}}{CH} \cdot COO(CH_2)_3CH_3} + CF_3COOH$$

It should be noted that tryptophan is unavoidably destroyed during acid hydrolysis. This amino acid has to be estimated by one of several other assay procedures, which will not be treated here.

A. EQUIPMENT AND SUPPLIES

Gas chromatograph as in Experiment 31

Column. Purchase custom-made from a gas chromatography supply house or prepare as follows:

1. Two columns with 0.325% EGA. Place 29.9 g of 80–100 mesh acid-washed H.T. Chromosorb in a round-bottomed flask (500 ml) and add acetonitrile until the Chromosorb is well covered. Add 0.098 g of EGA already dissolved in acetonitrile, mix by swirling, and evaporate the solvent using the rotary evaporator and vacuum while the flask is held in a water bath of approximately 60 C. Pack two glass columns, 1.5 m × 4 mm i.d., while gently tapping after each addition of 1–2 in. of material. Place a small plug of glass wool in each end and cure the columns overnight at 210 C with a slow stream of carrier gas while the exit ends of the columns are left disconnected.
2. Two columns with 1.5% OV-17. Place 30 g of 80–100 mesh H.P. Chromosorb G in the flask and cover with methylene chloride. Add

0.45 g of OV-17 already dissolved in the solvent and mix by swirling. Dry, pack, and cure the columns as instructed above, except that an oven temperature of 220 C is to be used.

Ultrasonic bath as described in Experiment 31

Controlled-temperature oil bath; all-glass container on top of hot plate–magnetic stirrer combination (Corning is satisfactory)

Oxygen torch

Nitrogen gas cylinder

Fritted glass filtration assembly, 20 mm diameter

Glass fiber filter pads (No. 934-H, Reeve Angel)

Freeze-dried potato tuber

Rotary vacuum evaporator with Teflon or glass surfaces only

10% TCA (trichloroacetic acid)

80% Ethanol

Methylene chloride

TFAA (trifluoroacetic acid anhydride)

Extractant (solution A): prepare 500 ml of a 0.04 *M* K_2HPO_4–0.015 *M* Na_2SO_3 solution

6 *N* HCl (solution B)

0.1 *N* HCl (solution C)

Internal standard solution (solution D): prepare 200 ml of 2.5 m*M* ornithine in 0.1 *N* HCl

3 *M* HCl (anhydrous) in butanol (solution E): follow directions for preparation of HCl-methanol (solution B) as given in Experiment 31

Standard solution (solution F): prepare 200 ml of 0.1 *N* HCl which is 2 m*M* for ornithine and 2 m*M* for each of the twenty protein amino acids except tryptophan, histidine, and hydroxyproline

B. ISOLATION OF PROTEIN

Grind approximately 2 g of freeze-dried potato tuber in a mortar. Transfer to a 50-ml beaker, add 20 ml of solution A, and stir at room temperature for 30 min with a magnetic stirrer. Centrifuge the slurry and decant and measure the volume of the clear supernatant solution. Add, while stirring, 2 volumes of 10% TCA. Centrifuge and discard the supernatant. Wash the precipitated protein twice with 80% ethanol and once with acetone, centrifuging and decanting after each wash. Remove the protein and dry in a vacuum oven at 40 C.

C. HYDROLYSIS OF THE PROTEIN

Weigh accurately 10–30 mg of the dried protein and transfer to an 18 × 220 mm test tube with 15 ml of solution B. Seal the tube in an oxygen flame while flushing the tube with nitrogen. Hydrolyze at 110 C in an oven for 20 hr. Filter the resulting amino acid solution on a fritted glass filter and collect the filtrate quantitatively. Transfer to a 100-ml round-bottomed flask with ground joint and bring to dryness using a rotary evaporator (no metal surfaces) at 50 C. Add 4 ml of solution C and 4 ml of solution D. Again bring to dryness under same conditions as before. Add 10 ml of methylene chloride and remove traces of moisture by azeotropic distillation using the evaporator. Repeat once.

D. FORMATION OF DERIVATIVES

Add 1.5 ml/mg protein of solution E and sonicate for 75 sec in an ultrasonic cleaner to disperse or dissolve the amino acids. Attach a drying tube filled with Drierite to the flask and heat for 15 min at 100 C in an oil bath to complete the esterification. Remove the solvent with the rotary evaporator as before. Add 2 ml of methylene chloride and 2 ml of TFAA (trifluoroacetic anhydride). Mix well and pour 2 ml of the mixture into an acylation tube. Flush with N_2, cap securely, and heat for 1 hr at 100 C in an oil bath. Do this in a ventilated hood behind a protective screen (plastic or shatterproof glass).

E. STANDARDS

Bring 200 ml of solution F to dryness under vacuum and further dry with methylene chloride as outlined above. Proceed as in Section D.

F. GAS CHROMATOGRAPHY

Inject in two separate runs 5 μl of the sample and standard derivatives with the EGA columns installed. Program the oven temperature at 4°/min with a 6-min delay, starting at 70 C, and with 180 C final temperature. Change to OV-17 columns and repeat the injections. Program at 5°/min, starting at 120 and ending at 210 C.

G. CALCULATIONS

From the peak areas of the standard chromatogram calculate the RMR values (relative molar response) of each of the amino acids:

$$\mathrm{RMR}_{\mathrm{a.a.}} = \frac{A_{\mathrm{a.a.}}(\mathrm{St.})}{A_{\mathrm{ornithine}}(\mathrm{St.})}$$

Use these values to calculate the following one, employing the peak areas obtained from the sample chromatograms:

$$\mathrm{mg}_{\mathrm{a.a.}} = (A_{\mathrm{a.a.}}/A_{\mathrm{orn}}) \times (\mathrm{MW}_{\mathrm{a.a.}}/\mathrm{MW}_{\mathrm{orn}}) \times \left(\frac{\mathrm{mg}_{\mathrm{orn}}}{\mathrm{RMR}_{\mathrm{a.a.}}}\right)$$

TABLE 17

EGA Column		*OV-17 Column*	
Amino Acid	*RMR*	*Amino Acid*	*RMR*
Ala	0.82	Series of unseparated peaks	
Val	1.04		
Gly	0.65		
Ile	1.06	Asp	—
Leu	1.19	Phe	—
Pro	0.97	Orn (int. st.)	—
Thr	0.97	Tyr	1.52
Ser	0.98	Glu	—
CySH	0.74	Lys	—
Met	1.04	Arg	0.91
Phe	1.59	Cys	1.27
Glu	1.41		
Tyr			
Orn (int. st.)			
Lys	1.11		

Percentage of amino acid on a weight basis in proteins:

$$W/W\%_{\text{oa.a.}} = \frac{\text{mg}_{\text{a.a.}}}{\text{mg}_{\text{protein}}} \times 100$$

Use Table 17 to identify the individual amino acids. The amino acids emerge from the columns in the order listed. The RMR values obtained from the standard should be close to the values shown in the table.

REFERENCE

1. Gehrke, C. W., D. Roach, R. W. Zumwalt, D. L. Stalling and L. L. Wall. *Quantitative Gas-Liquid Chromatography of Amino Acids in Proteins and Biological Substances*. Analyt. Biochem. Lab., Columbia, Missouri. 1968.

EXPERIMENT

34 COMPOSITION OF THE FREE AMINO ACID POOL OF PLANT TISSUES

In contrast to animal tissues, a relatively large portion of the total so-called crude protein in plants exists, not as true protein, but as free amino acids. This portion often exceeds 50% of the total. The size and composition of this amino acid pool are known to fluctuate within wide limits depending on nutritional status, stage of growth, and many other factors. Information leading to the characterization of the pool is therefore important in identifying and studying such factors as well as in recognizing limiting factors in protein synthesis and in evaluating food materials in terms of human and animal nutrition.

In this experiment the free amino acids are extracted and purified by a "clean-up" procedure to eliminate interfering substances such as free sugars and minerals. The purified amino acids are subsequently prepared for gas chromatography by the same procedures as in Experiment 33. It is important to realize that tryptophan is not destroyed in this analysis, as it was in the previous experiment, since acid hydrolysis is not utilized.

The amino acids which were not observed in Experiment 33 will appear in the chromatograms. One of them is pyrollidone-5-carboxylic acid (PCA, also called pyroglutamic acid), which is produced during sample preparation from glutamine present in the sample. The other is γ-aminobutyric acid (Gaba), which does not occur in protein but is frequently found at relatively high concentrations in plant materials as a free acid.

A. EQUIPMENT AND SUPPLIES

Gas chromatograph as in Experiment 31
Columns as in Experiment 33
Rotary vacuum evaporator with Teflon or glass surfaces only
Cation exchange resin (Amberlite IR-120H)
Ion exchange column; 25-ml burette with a plug of glass wool
Controlled-temperature oil bath as in Experiment 33
Freeze-dried potato powder
Extractant (solution A): prepare a solution containing 350 ml of absolute ethanol, 5 ml of concentrated hydrochloric acid, 66 mg (0.5 mmole) of ornithine, and water to make 500 ml
0.1 *N* HCl (solution B)
7 *N* NH_4OH (solution C).
3 *M* Anhydrous HCl in butanol (solution D): see solution E, Experiment 33
Standard solution (solution E): prepare 200 ml of 0.1 *N* HCl which is 205 m*M* for ornithine and for each of the twenty amino acids that occur in proteins except histidine and hydroxyproline but including tryptophan and Gaba

B. EXTRACTION AND DERIVATIVE FORMATION

Add 20.00 ml of solution A to 1.00 g of freeze-dried potato powder and shake for 1 hr. Filter rapidly through coarse filter paper and transfer 5.00 ml to a 100-ml round-bottomed flask with a standard taper joint. Evaporate to dryness under vacuum at 30–40 C with a rotary evaporator and dissolve the residue in 8 ml of solution B. Pass the solution slowly (20 drops/min) through a 3.25-in. column of Amberlite IR-120 H (use a 25-ml burette). Wash the resin twice with 20 ml of glass-distilled water at a rate of 3 ml/min and discard the washings. Add in sequence two 6-ml portions of solution C to the column, elute the amino acids at a rate of 20 drops/min, and follow with two 5-ml portions of water. Collect the eluate and washings in a 100-ml flask, bring to dryness under vacuum, and dry azeotropically with 2 × 10 ml of methylene chloride (see Experiment 33). Add 15 ml of solution D; proceed as indicated in Experiment 33, Section D, and sonicate.

C. STANDARDS

Add 2.00 ml of solution E to 4.7 ml of ethanol in a 100-ml flask and evaporate to dryness with the rotary evaporator. Proceed as in Section B, starting with "and dissolve the residue."

TABLE 18

EGA		*OV-17*	
Amino Acid	*RMR*	*Amino Acid*	*RMR*
Ala	0.93	Series of unseparated peaks	
Val	1.25		
Gly	0.75		
Ile	1.18	Asp	—
Leu	1.26	Phe	—
Pro	1.15	Orn (int. st.)	—
Thr	1.02		
Ser	0.87	Tyr	1.24
Gaba	0.71	Glu	—
PCA	1.12	Lys	—
Met	1.02	Arg	0.61
Phe	1.43	Trp	—
Asp	1.28	Cys	—
Glu	1.39		
Tyr			
Orn (int. st.)	—		
Lys	1.09		
Trp	0.84		

D. GAS CHROMATOGRAPHY AND CALCULATIONS

These are identical with those in Experiment 33 with the following exceptions. On the EGA column, Gaba is eluted after serine and is followed by PCA, which appears in the position normally occupied by cysteine. (Cysteine, if present in the tissue, will be oxidized to cystine during the clean-up operations and will appear on the OV-17 chromatogram). Owing to unequal losses during clean-up, the RMR values will be different from those used in Experiment 33. The values in Table 18 are typical and should serve as a guide in evaluation of technical adequacy.

EXPERIMENT 35 ASSAYS OF PECTIN METHYLESTERASE, FREE METHANOL, AND METHOXY GROUPS IN PLANT TISSUES

Pectin methylesterase (PME) is probably universally present in living plant tissues and is known, or suspected, to play important roles in processes such as abscission, cell wall extension, and fruit ripening. Study of its precise function in these processes has been hampered by lack of suitable assay methods. This experiment introduces sensitive and precise assay methods that should allow rapid progress in this area.

Methanol, which is generated by the enzyme (PME) from a pectin solution, released from tissue pectin by alkali, or exists free in the tissue as the result of prior enzyme activity, is converted into methyl nitrite by letting the sample react with a nitriting reagent. Methyl nitrite is assayed by GLC.

$$\text{Pectin} \xrightarrow{\text{PME}} \text{pectinic acid} + \text{MeOH}$$

$$\text{MeOH} + \text{HNO}_2 \longrightarrow \text{MeNO}_2 + \text{H}_2\text{O}$$

A. EQUIPMENT AND SUPPLIES

7% H_3PO_4 in water (solution A)
5% KNO_2 in water (solution B)
Nitrating solution (solution C): mix equal volumes (2×50 ml) of the *cold* solutions A and B; use within 1 hr.
1 *M* K_2HPO_4, pH 7.5 (solution D)
0.5 *N* NaOH with 0.25% EDTA in water (solution E)

FIGURE 64. A nitrating tube.

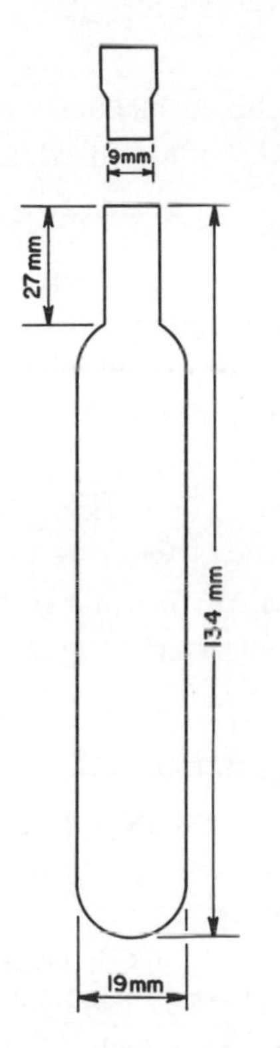

1% Pectin in 0.2% K_2HPO_4, pH 7.5 (solution F). Weigh 10.0 g of citrus pectin (Sunkist Growers) into a 50-ml beaker. Moisten with a few milliliters of glycerol and stir to make a uniform slurry. Add approximately 30 ml of water and stir to make a uniform suspension. Pour the suspension into a 2-liter flask and add water to make 1 liter. Gently heat while stirring until the solution comes to a boil. Cool and store in a refrigerator.

Propanol standards: prepare 1 liter each of 0.01% by volume (solution G) and 0.001% (solution H) of *n*-propanol

Standard methanol solutions (solutions I_1–I_{10})

1. Prepare 100 ml each of 0.001, 0.002, 0.004, 0.006, and 0.008% methanol in H_2O; each solution should also contain 0.005% *n*-propanol
2. Prepare five similar solutions containing 0.01, 0.02, 0.04, and 0.05% methanol, respectively, and 0.02% *n*-propanol each

Tomatoes

Nitrating tubes (Figure 64)

B. PROCEDURE

1. Sample Preparation

Grind 20 g of cold tomato tissue in a precooled blender with 20 ml of cold solution D (ice temperature) until well homogenized.

2. Determination of Free Methanol

Add 5.0 ml of the slurry to a stoppered, prechilled test tube containing 5.0 ml of solution H. Mix well and immediately transfer 1.0-ml portions to duplicate nitrating tubes (Figure 64) containing 5 ml of solution C. Cap with serum caps, immerse the tubes to the "shoulder" in crushed ice, and mix the contents well by shaking at intervals. Standardize the number of times and manner of shaking. Remove approximately 1 ml of the head space gas (head space is the empty space above the liquid within the tube) with a hypodermic syringe and inject into the gas chromatograph.

3. PME Assay

Transfer 9.0 ml of solution F to four stoppered test tubes and place in a 30 C water bath. Allow sufficient time for the pectin solution to reach 30 C and add 1.0-ml portions of the tomato slurry to tubes at convenient (5 min) intervals. Mix and allow the enzyme to react for exactly 3 min. Omit the reaction time for the two blank determinations. Transfer 1.0 ml of the mixture to nitrating tubes (see 2 above, "Determination of Free Methanol") in an ice bath. Sample the head space and inject into the gas chromatograph as before.

4. Determination of Methoxy Groups

Transfer 10 ml of the tissue slurry to 60-ml test tubes containing 10 ml of solution G and add 20 ml of solution E. Stopper securely and shake for at least 30 min at room temperature. Transfer 1.0-ml portions to nitrating tubes and proceed as before.

C. STANDARD CURVES

Transfer 1.0-ml portions of each of the solutions I to nitriting tubes and proceed with GLC as before. Measure peak heights and compute the relative weight response (RWR) as indicated under "E, Calculations" below.

D. GAS CHROMATOGRAPHY

A gas chromatograph with either single or dual column systems and equipped with flame ionization detection may be used. The separation of components in the head space gas is easily obtained with 6 ft × 0.25

in. (i.d.) columns of stainless steel or glass and packed with a low-polarity liquid phase on Chromosorb W. A column packed with 15% Ucon Non-Polar LB1715 on 60–70 mesh Chromosorb W is adequate when operated at 40–50 C with a carrier gas flow rate of 50–60 ml/min.

E. CALCULATIONS

Obtain the RWR of methanol with respect to *n*-propanol using peak height measurements from the standard solutions. Then, for free methanol and total methoxy (as methanol):

$$\%\text{MeOH} = \frac{P_{\text{MeOH}} \times \%\text{PrOH}}{P_{\text{PrOH}} \times \text{RWR} \times D} = \text{percent MeOH in tissue}$$

where P_{MeOH} = peak height of methanol in the sample
P_{PrOH} = peak height of propanol in the sample
% PrOH = known concentration of propanol in the sample solution

$$\text{RWR} = \text{relative weight response} = \frac{P_{\text{MeOH}}/\%\ \text{MeOH}}{P_{\text{PrOH}}/\%\ \text{PrOH}}$$

D = dilution factor of tissue in sample solution (g/ml)

For PME activity determinations obtain

$$\Delta\text{MeOH} = \%\ \text{MeOH}_t - \%\ \text{MeOH}_z$$

where t and z refer to incubation times of t (usually 3) and zero (blank) minutes, respectively. Then

$$\text{PME activity} = \frac{\Delta\text{MeOH} \times 10^4}{t \times 32}\ \mu M/\text{min/g tissue}$$

APPENDICES

APPENDIX

1 TABLE FOR ADJUSTMENT OF CONCENTRATION OF AMMONIUM SULFATE SOLUTIONS[a]

	% Saturation																	
	10	20	25	30	35	40	45	50	55	60	65	70	75	80	85	90	95	100
$(NH_4)_2SO_4$, g/l to bring from 0%	55	113	144	175	209	242	278	312	350	390	430	474	519	560	608	657	708	760
$(NH_4)_2SO_4$, g to make up 1 liter of solution	53	106	133	160	187	214	240	266	294	321	345	380	404	429	456	482	510	536
$(NH_4)_2SO_4$, g/l added to bring from % saturation																		
10		57	87	118	149	182	215	250	287	325	365	405	448	494	530	585	634	685
20			29	59	90	121	154	188	225	260	298	337	379	420	465	512	559	610
25				29	60	91	123	157	192	228	265	304	345	386	430	475	521	571
30					30	61	93	125	160	195	232	270	310	351	394	439	485	533
35						30	62	94	128	163	199	235	275	315	358	403	449	495
40							31	63	96	131	166	205	240	280	322	365	410	458
45								31	64	98	133	169	206	245	286	330	373	420
50									32	65	100	135	172	211	250	292	335	380
55										33	66	101	138	176	214	255	298	344
60											33	67	103	140	179	219	261	305
65												34	69	105	143	182	224	267
70													34	70	108	146	187	228
75														35	72	110	149	190
80															36	73	112	152
85																37	75	114
90																	37	76
95																		38

[a] The values are for 25 C.

APPENDIX

2 PROTEIN ESTIMATION BY ULTRAVIOLET LIGHT ABSORPTION

Ratio 280/260	*% Nucleic Acid*	*B 280*	*Factor for 0.5-cm cell*[a]	*1.0-cm cell*[a]
1.75	0	2.08	2.21	1.11
1.60	0.25	2.14	2.15	1.08
1.50	0.50	2.19	2.10	1.05
1.40	0.75	2.25	2.05	1.02
1.30	1.00	2.31	1.99	1.005
1.25	1.25	2.36	1.95	1.000
1.20	1.50	2.42	1.90	0.950
1.15	2.00	2.53	1.82	0.901
1.10	2.50	2.65	1.74	0.870
1.05	3.00	2.76	1.67	0.837
1.00	3.50	2.87	1.60	0.804
0.96	3.75	2.93	1.57	0.773
0.92	4.25	3.05	1.51	0.757
0.88	5.00	3.22	1.43	0.717
0.86	5.25	3.27	1.41	0.705
0.84	5.50	3.33	1.38	0.690
0.82	6.00	3.44	1.34	0.670
0.80	6.50	3.56	1.29	0.649
0.78	7.25	3.73	1.23	0.619
0.76	8.00	3.90	1.18	0.597
0.74	8.75	4.07	1.13	0.567
0.72	9.50	4.24	1.08	0.544
0.70	10.75	4.52	1.02	0.510
0.68	12.00	4.80	0.96	0.481
0.66	13.50	5.14	0.90	0.449
0.65	14.50	5.37	0.86	0.429
0.64	15.25	5.54	0.83	0.417
0.62	17.50	6.05	0.76	0.382
0.60	20.00	6.62	0.70	0.361
0.49	100.00	—	—	—

[a] Factor × 280 reading = mg/cc.

APPENDIX

3 PROPORTION OF ^{32}P ACTIVITY REMAINING AFTER X DAYS

$$\log_{10} \frac{N}{N_0} = -0.02105t \quad (t \text{ in days})$$

t, days	*antilog −0.02105t*	*t, days*	*antilog −0.02105t*
1	0.953	25	0.298
2	0.908	26	0.283
3	0.865	27	0.270
4	0.824	28	0.257
5	0.785	29	0.245
6	0.748	30	0.234
7	0.712	31	0.223
8	0.679	32	0.212
9	0.646	33	0.202
10	0.616	34	0.192
11	0.587	35	0.183
12	0.559	36	0.175
13	0.533	37	0.166
14	0.507	38	0.159
15	0.483	39	0.151
16	0.461	40	0.144
17	0.439	41	0.137
18	0.418	42	0.131
19	0.398	43	0.124
20	0.379	44	0.119
21	0.361	45	0.113
22	0.344	46	0.108
23	0.328	47	0.102
24	0.312	48	0.098

APPENDIX

4 SPECIAL PROCEDURES FOR DISC ELECTROPHORESIS*

A. INSTRUCTIONS FOR DESALTING (DIALYZING) SAMPLES FOR ELECTROPHORESIS

Dialysis is desirable for materials which have been prepurified by ammonium sulfate treatment.

B. INSTRUCTIONS FOR STAINING WITH P.A.S. (PERIODIC ACID, SCHIFF REAGENT)

This procedure is used on glycoproteins, polysaccharides, chondroitin sulfate, heparin, and other high-carbohydrate materials.

1. Prepare a 0.2% aqueous solution of periodic acid (e.g., 100 mg in 50 ml of distilled water).

2. After electrophoresing the sample in the usual way, place the gel column in 7.5% acetic acid at room temperature for 1 hr. (Small plastic centrifuge tubes with several holes drilled in the side wall and equipped with corks are useful vessels for holding gels during these procedures.

3. Place all gels in the P.A. solution for 1 hr.

4. Remove the P.A. electrophoretically (using 7.5% acetic acid in the cold for 1 hr).

5. Place the gel column in the Schiff reagent (prepared as described in Section C, 1, "Feulgen Reagent Solution" below) and keep in the cold until the bands appear red. The gels may be placed in stoppered test tubes containing Schiff reagent without loss of stain for about 1 month at room temperature.

6. If long-term retention of stain is desired, prepare two working solutions: (a) 1 N HCl; (b) 10% aqueous potassium bisulfite ($KHSO_3$) or 10% aqueous potassium metabisulfite ($K_2S_2O_5$). Mix equal parts of these solutions with 10 parts of water. Place the stained gel column in stoppered test tubes containing this solution for at least 2 hr at room temperature. At the end of 2 hr (or longer period) replace with fresh solution and repeat seven times at 2-hr intervals. Finally, store the sample in the stoppered tubes containing this rinse solution.

* This procedure was adapted from a method of Canal Industrial Corporation, 4935 Cordell Avenue, Bethesda, Maryland.

C. INSTRUCTIONS FOR FEULGEN STAINING OF DNA IN GELS

1. Feulgen Reagent Solution (Schiff)

Add 0.5 g of basic fuchsin or pararosaniline to 100 ml of water at room temperature. Add 1 g of potassium (or sodium) metabisulfite and 10 ml of 1 *N* HCl. Shake at intervals until the solution is straw-colored (about 3 hr). Add 0.25–0.5 g of activated charcoal, shake, filter, and store the solution in a tightly stoppered bottle in the refrigerator. The reagent should be water-clear; if it is still yellow, add more charcoal and filter.

2. Rinse Solution

The rinse solution is prepared by mixing 10 ml of 1 *N* HCl, 10 ml of 5% potassium (or sodium metabisulfite) and 180 ml of H_2O.

3. Procedure

(a) After electrophoresis, remove the gel from the tube and place it in ice-cold 1 *N* HCl (in the cold) for about 0.5 hr. Use 20 ml HCl/gel

(b) Promptly place the gel in a solution of 1 *N* HCl which has been preheated and maintained at 60 C for 12 min.

(c) Place the gel in the Feulgen reagent solution in a closed container at room temperature for about 1 hr or until the reagent has penetrated the most central portion of the disc.

(d) Store the gel in the Feulgen reagent (good for about 1 month).

(e) For permanent storage, rinse the gel in the rinse solution (7 changes of 2 hr each) and store in the rinse solution.

APPENDIX

5 METHOD OF DISC ELECTROPHORESIS OF PROTEINS

A. PROCEDURE

1. Preparation of Protein Sample

Grind 50 g of tissue in an equal amount (w/v) of cold 0.01 *M* Tris buffer, pH 7.5, with a high-speed homogenizer (ice temperature) for alternate half-minute periods at high and low speeds (for a total of 4.5 min). Strain the homogenate through two layers of cheesecloth and centrifuge at 10,000 × *g* for 30 min in the cold. The supernatant is brought to 30% saturation with solid $(NH_4)_2SO_4$ while being slowly stirred in an ice bath. Centrifuge at 5000 × *g* for 10 min and save the supernatant. Add solid $(NH_4)_2SO_4$ to the solution to give a final $(NH_4)_2SO_4$ concentration of 70% saturation, and centrifuge again. Suspend the pellet in 5 ml of 0.01 *M* Tris buffer, pH 8.35, and dialyze against the same buffer in the cold for 36 hr. The dialyzed material should be diluted with buffer to give a solution containing approximately 200–400 μg protein/0.1 ml. Protein concentration may be determined by the Folin method (see Experiment 1).

2. Electrophoresis

(a) Stock Solutions (1).

A

1 *N* HCl	48. ml
Tris	36.6 g
TEMED	0.46 ml
H_2O	to 100 ml

(pH 8.9)

B

1 *M* H_3PO_4	25.6 ml
Tris	5.7 g
H_2O	to 100 ml

(pH 6.9)

C

Acrylamide	30.0 g
Bis	0.8 g
$K_3Fe(CN)_6$	15.0 ml
H_2O	to 100 ml

D

Acrylamide	10.0 g
Bis	2.5 g
H_2O	to 100 ml

E

Riboflavin	4.0 mg
H_2O	to 100 ml

(b) Buffer. Tris, 6.0 g; glycine, 28.8 g; brought to 1 liter, pH 8.3.
Working solutions:
Lower gel 1: 1 part A; 2 parts C; 1 part H_2O
Lower gel 2: 0.14 g $(NH_4)_2S_2O_8$; H_2O to 100 ml
Sample gel: 1 part B; 2 parts D; 1 part E
Spacer gel: 1 part sample gel; 1 part H_2O

Note: All stock working solutions should be kept in brown bottles and stored in the cold. Lower gel No. 2 is stable for a little over a week.

Incubate lower gel solutions 1 and 2 in separate tubes at 35–40 C for 10 min and mix 1:1. Fill glass tubes (5 × 63 mm) to height of 5 cm (the bottoms of the tubes are stoppered). Carefully layer 3–4 mm of water above the gel and place in a 37 C oven to allow gels to polymerize (approximately 20 min). Carefully shake out water and add a drop of spacer gel to each tube and then remove the spacer gel. Add 1 cm of spacer gel and carefully layer 3–4 mm of water above it. Place the samples under a fluorescent lamp until polymerized (15–20 min). Shake out the water and rinse with the spacer gel. Then layer a mixture of sample gel and sample, 0.1 ml of each, over the spacer gel, and allow this to polymerize under a fluorescent lamp. Remove the stoppers and place the tubes in an electrophoresis chamber. Add buffer (Tris-glycine, pH 8.3) to the upper and lower chambers. (The upper chamber contains the cathode, and the lower one the anode.) The upper buffer also contains 0.05 mg of bromphenol blue in 200 ml; the dye will be used to follow progress of the front. The power supply is connected and the current adjusted to 3 mamp/tube. Electrophoresis is carried out until free bromphenol blue dye has migrated about 4.5 cm into the lower gel (separation gel). After removal of the tubes from the electrode chambers, the gel is extruded into tubes containing 0.5% Amido Schwartz in 7.5% acetic acid and allowed to stand overnight or a minimum of 2 hr. (Rimming the gel with a blunt needle under water aids in its removal.) Excess staining solution is decanted and the gel (sample gel uppermost) is placed in destaining tubes (electrophoresis tubes). Electrophoretic destaining is performed in the same apparatus, both the upper and the lower chambers containing 7.5% acetic acid. The power supply is connected and the current adjusted to 10 mamp/tube. Destaining should take about 2 hr. Finally the destained gel can be placed in 7.5% acetic acid for storage.

For alternative procedures for gel preparation see (2) and (3). Those presented in (3) are useful when a sample gel is not desired or when the protein sample may contain inhibitory substances that prevent gel formation.

REFERENCES

1. Chemical Formulation for Disc Electrophoresis. Canaico publication.
2. Davis, B. J. Disc Electrophoresis. II. Method and Application to Human Serum Proteins. *Ann. N.Y. Acad. Sci.* 121: 404 (1964).
3. Steward, F. C., R. F. Lyndon and J. T. Barber. Acrylamide Gel Electrophoresis of Soluble Plant Proteins: A Study on Pea Seedlings in Relation to Development. *Am. J. Bot.* 52: 155 (1965).

APPENDIX 6 METHOD OF CONVERTING DOWEX-1 FROM THE CHLORIDE INTO THE FORMATE FORM

The Dowex-1 should be washed several times with distilled water and the fine and coarse particles removed by sedimentation. The Dowex is then placed in a large glass column containing a fritted disc padded with a layer of glass wool. The Dowex is washed with 3.0 *M* sodium formate until no more chloride ions are removed. (An $AgNO_3$ spot test may be used to detect Cl^-.) The Dowex is further washed with three bed volumes of a mixture of 6.0 *N* formic acid and 1.0 *M* sodium formate, then with several bed volumes of 88% formic acid. It is finally washed with deionized water until the effluent is of neutral pH.

APPENDIX 7 PREPARATION OF A STANDARD BENTONITE SUSPENSION (1)

Suspend 10 g of technical grade bentonite in 200 ml of water by thorough stirring. Remove and discard the coarse particles by centrifuging at 4000 × *g* for 15 min. The fine clay suspension is decanted and centrifuged at 9000 × *g* for 20 min. The cloudy supernatant liquid is discarded and the sediment is suspended in 0.01 *M* acetate buffer, pH 6.0. After thorough stirring, the suspension is centrifuged at 9000 × *g* for 15 min and the supernatant liquid is discarded. Repeatedly suspend the clay pellet in acetate buffer and centrifuge at 20,000 × *g* for 15 min until at 260 mμ the absorbancy of the supernatant liquid in a 1-cm cell is less than 0.7. The bentonite is finally suspended in acetate buffer to yield a concentration of 40 mg/ml (determine the concentration by drying a 1- to 2-ml sample of the bentonite suspension).

REFERENCE

1. Brownhill, T. J., A. S. Jones and M. Stacey. *Biochem. J.* 73: 434 (1959).

APPENDIX
8 PREPARATION OF CELLULOSIC ABSORBENTS FOR ION EXCHANGE CHROMATOGRAPHY (1)

A. INTRODUCTION

Ion exchange chromatography, particularly that employing substituted celluloses, has proved to be a valuable tool for purification of polyionic biological molecules such as proteins. Though many substituted celluloses are available, the two most widely employed are the anion exchanger diethylaminoethyl (DEAE) and carboxymethyl (CM) celluloses. This discussion will center on the preparation and equilibration of these two materials for column chromatography.

1. Precycling

Stir the ion exchanger into 15 volumes (i.e., vol liquor/dry weight of ion exchanger) of "first treatment" acid or alkali (see below) and leave for at least 30 min. Filter or decant off supernatant liquor and wash in a funnel until the effluent is at the "Intermediate pH" given in the accompanying tabulation. Stir the ion exchanger into 15 volumes of "second treatment" acid or alkali and leave for another 30 min. Repeat the second treatment and follow by a wash in a funnel until the filtered effluent is near neutral.

(a) Order of Treatment.

	First Treatment	*Intermediate pH*	*Second Treatment*
DEAE	0.4 *N* HCl	4	0.5 *N* NaOH
CMC	0.5 *N* NaOH +0.5 *M* NaCl	8	0.5 *N* HCl

2. Degassing of DEAE

Place the DEAE cellulose in the acid component of the buffer with the pH below 4.5. If the concentration of the acid component of the buffer is not high enough to give pH 4.5, a higher concentration of the acid component must be used. Apply a good vacuum (down to below 10 cm of Hg pressure) with stirring, until no more bubbles are noticed but before boiling occurs. This may conveniently be carried out by stirring the slurry with a magnetic stirrer in a stoppered Buchner flask connected to a good water pump. A water trap should be placed between the flask and the pump. Add the basic component of the buffer to give the desired pH. Equilibrate the cellulose as outlined below. Ideally, for more critical work, buffer solutions must be made up from CO_2-free water and kept free of CO_2.

3. Removal of Fines

Fines, small particles of cellulose exchanger, can be used in column chromatography but tend to decrease the flow rates and cause uneven resolution and should be removed. To prevent generation of fines, slurries of cellulose should not be shaken or stirred vigorously. Disperse the ion exchanger in the buffer. The total volume of the slurry should be 30 ml/g of dry ion exchanger used or about 6 ml/g of wet filtered ion exchanger. Allow the slurry to settle in a suitable measuring cylinder in an area free of drafts, direct sunlight, heaters, etc. The time allowed for settling is calculated from the equation

$$t = nh$$

where

t = time, min

h = the total height of the slurry in the measuring cylinder, cm

n = a factor between 1.3 and 2.4

The choice of a suitable value of n depends upon the degree of fines removal required. It might not be necessary to use an n value less than 1.3 to achieve the desired flow characteristics. When $n = 2.4$, only the finest particles will be removed. Note the "wet settled volume," the volume occupied by the ion exchanger after settling under prescribed conditions. Immediately remove the supernatant buffer solution containing fines so that the final volume remaining in the measuring cylinder is the "wet settled volume" plus 20%. Make up the slurry with buffer so that the final volume of the slurry is 150% of the "wet settled volume" of the ion exchanger.

4. Equilibration

Equilibration is one of the most important steps in cellulose ion exchange preparation. Improper equilibration often leads to irreproducible results, particularly when one is working with dilute buffers in biological purifications. For the equilibration of all ion exchange celluloses one of the following methods may be used.

(a) Aliquot Buffer Changes. Stir the ion exchanger into a volume of the buffer (15–30 mg/dry g). Leave for 10 min and decant or filter off the supernatant liquor. Repeat the treatment until the *filtrate* of the supernatant liquor has *exactly* the same pH and conductivity as the buffer. This must be checked after a further buffer change. This method may require many changes and may be time-consuming when buffers of low concentrations are used.

(b) pH Adjustment Followed by Buffer Changes. Stir the ion exchanger into a volume of the buffer (15–30 ml/dry g). Titrate with the acid (for DEAE cellulose) or the base (for CM cellulose) component of the buffer to the correct pH. Decant or filter off the supernatant liquor. Carry out aliquot buffer changes as described in method (a) above. This method is suitable for most buffers.

(c) Equilibration with Concentrated Buffer Followed by Re-equilibration in the Column with the Starting Buffer. Equilibrate the ion exchanger with a more concentrated (3–10 times) buffer of the same pH as the starting buffer by either method (a) or (b) above. Pour the column (see below). Pass the starting buffer through the column until the conductivity and the pH of the effluent are *exactly* the same as those of the starting buffer. This method is suitable for column separations starting with low concentrations of buffer. It must be emphasized that readings of pH and conductivity *must* be exact. With true equilibrium, the equilibrating solution will be identical with the starting buffer solution. It is essential that readings of two consecutive equilibration solutions be not only identical to each other but also with the starting buffer.

B. COLUMN TECHNIQUES

1. Column Packing

Convection currents in the slurry must be avoided at all costs during the actual column packing. Set up the column vertically in an area free of drafts, direct sunlight, heaters, etc. Pour the stirred slurry into the column. Allow the effluent from the column to run to waste. When all the slurry is added, attach or insert the top column end. Pump or run the buffer through the column at a flow rate of at least 45 ml/hr/cm^2 of the internal cross-sectional area of the column (ml/hr/cm^2) until the column bed height is constant.

It is essential that, from the moment of pouring the slurry into the column to the stage of having a settle column bed of ion exchanger, the operations be carried out as quickly as possible; otherwise convection currents in the slurry have sufficient time to be set in motion. Finally, stop the flow of buffer into and out of the column.

2. Introduction of Sample

Dissolve or equilibrate the sample in the starting buffer. It may be necessary to carry out some adjustment of the pH. Always load the mixture to be separated at a controlled flow rate. If there is not available a column with a piston-type end unit which allows pumping the sample on, the sample may be layered on the bed surface with a pipet. If this is done, the top of the bed should be protected with a porous material.

3. Elution

Start the elution immediately or at a standardized time after the sample has been added.

Generally, there are three methods by which chromatographic separations are achieved.

(a) Starting Buffer Elution. The same buffer is used for the equilibration of the ion exchanger and the mixture and for the separation.

The starting buffer elution may be used in two different ways.

(1) For the removal of all the components from the mixture except the desired component, a short fat column is generally used. The amount of ion exchanger required would depend on its capacity for the unwanted components in the mixture. The wanted component should appear right after the void volume of the column.

(2) Starting buffer development. This is "true" chromatography and may be used when the mixture consists of chromatographically similar components. A relatively long column is normally required for this type of separation in order to obtain the optimum resolution of the different components in the mixture. It is advisable to use only a small part of the total capacity of the column.

(b) Gradient Elution. A buffer of a continuously changing composition is used to effect the separation. The variation in the composition of the buffer may be one from lower to higher ionic concentrations or of the appropriate pH or both. Since the buffer itself is the main factor in the achievement of the separation, the amount of ion exchanger required would depend on the capacity of the ion exchanger for the mixture. Should the mixture, however, contain any chromatographically similar components, some additional length to the column would be required in order to obtain resolution.

(c) Stepwise Elution. Different buffers are used for the equilibration of the ion exchanger and for the chromatographic separations on the column. This method should be avoided unless the system has been well defined, since the results may be misleading.

4. Reuse

If a clean elution is achieved, it is necessary only to reequilibrate the ion exchanger for reuse. However, if any material has been left on the column, it would have to be removed by either: (a) using stronger elution conditions; (b) removing the ion exchanger from the column and applying the precycling procedure; or (v) Physically removing the contaminated portion of the column bed and extruding the rest of the column for reuse.

The ion exchanger would, in all cases, have to be reequilibrated.

REFERENCE

1. Whatman's information Leaflet (IL2), Advanced Ion Exchange Celluloses. (All procedures dealing with cellulosic absorbents are based on this leaflet.)

APPENDIX

9 RADIOISOTOPES

A. THEORY

The nucleus of any atom is composed of protons and neutrons, the mass being relative to O_{16} (exactly 16.000). Electrons do not contribute significantly to the mass. Protons and neutrons have a mass number of 1; protons are positively charged, and neutrons have no charge.

The chemical properties of an atom are largely determined by the nuclear charge or number of protons in its nucleus, called the atomic number. Therefore the addition of neutrons to the nucleus changes the mass without changing the atomic number or nuclear charge and generally without changing the chemical properties.

The description of a nucleus usually includes the mass number, symbolized by a superscript A to the left of the element symbol.

$$\text{Mass number} \quad {}^{A}X_{Z} \quad \text{atomic number}$$

The atomic number or the number of unit positive charges (protons) carried by the nucleus is symbolized by a subscript Z to the right. Thus

${}^{1}H_{1}$	${}^{23}Na_{11}$	${}^{32}P_{15}$	${}^{14}C_{6}$
n = 0	12	17	8
p = 1	11	15	6

$A - Z =$ number of neutrons

Only certain combinations of neutrons and protons are stable. In stable nuclei the ratio n:p varies from 1 to 1.5. Elements whose nuclei vary in nuclear mass from the stable situation are called isotopes.

An excess of a neutron or proton relative to the stable condition leads to a spontaneous redistribution of particles to bring about a more stable condition of n and p. This stable condition may be the formation of an atom with different chemical properties, i.e., the formation of another element.

In discussing isotopes, certain nuclear particles must be defined:

(1) n: neutron

(2) p: proton

(3) d: deuteron, heavy hydrogen ${}^{2}H_{1}$

(4) α Particle: helium nucleus, $^4He_2$
(5) β^- Particle: negative electron
(6) β^+ Particle: positive, electron, does not exist in the nucleus
(7) γ Radiations: high-energy photons do not exist in the nucleus but are produced by nuclear transformations sometimes

When there is an unstable ratio of n to p, there are several kinds of nuclear transformations:

(1) Neutron $\rightarrow$ proton + β^- emission
(2) Proton $\rightarrow$ neutron + positron (β^+ emission) or protron + electron (captured) $\rightarrow$ neutron
(3) Emission of α particle:
2n + 2p $\rightarrow$ α particles ($^4He_2$)
(4) Other emissions might be X-rays or γ-rays from nuclear transformations.

Let us consider carbon, $^{12}C_6$: No more than 6 or 7 neutrons can be held in the nucleus with the 6 protons. The stable forms of C are $^{12}C_6$ and $^{13}C_6$; $^{14}C_6$ is unstable and transforms to $^{14}N_7$. How does it do this?

In considering what happens, note that a neutron is converted into a proton.

Number of	$^{14}C_6$	$^{14}N_7$
n	8	7
p	6	7

So, when $^{14}C_6 \rightarrow {}^{14}N_7$,

$$n \rightarrow p + \beta^-$$

When a nucleus spontaneously undergoes transformation, there is less energy in the stable nucleus than in the unstable form, that is, energy is lost. The energy involved is usually given in Mev units.

$$1 \text{ mass unit } (1/16 \text{ of O}) = 931 \text{ Mev}$$

The reaction for $^{14}C_6$ is then

$$^{14}C_6 \rightarrow {}^{14}N_7 + \beta^- + 0.15 \text{ Mev} \qquad \text{Half-life } (T_{1/2}) \; 5500 \text{ yr}$$

For ^3H,

$$^3H_1 \rightarrow {}^3He_2 + \beta^- + 0.015 \text{ Mev} \qquad 12.3 \text{ yr}$$

For ^{32}P,

$$^{32}P_{15} \rightarrow {}^{32}S_{16} + \beta^- + 1.72 \text{ Mev} \qquad 14.3 \text{ days}$$

In all of the reactions above, n $\rightarrow$ p + β^-.

The greater the energy (Mev), the greater is the penetration power of the particle emitted and usually the easier it is to measure.

B. CALCULATIONS INVOLVING ISOTOPES

Initially 1 curie was defined as the amount of radiation given off by 1 g of radium, which was 3.7×10^{10} disintegrations/sec (dps). Today a curie is defined as that amount of any isotope which emits 3.7×10^{10} particles/sec.

For any given element the rate of decay (t = time) is proportional to the number (N) of nuclei of the element present

$$\frac{dN}{dt} = \frac{\text{change in nuclei}}{\text{change in time}}$$

The rate of decay is expressed by the equation

$$-\frac{dN}{dt} = \lambda N$$

where λ = radioactivity decay constant. To calculate the amount of isotope after a given time of decay this equation can be changed to

$$\frac{dN}{N} = -\lambda\, dt$$

Then, in order to calculate the amount of isotope N after a certain time, the preceding equation is integrated:

$$\int_{N_0}^{N} \frac{dN}{N} = -\lambda \int_0^t dt$$

where N_0 = number of nuclei at time 0.

$$N = N_0 e^{-\lambda t}$$

or

$$\boxed{\ln \frac{N}{N_0} = -\lambda t}$$

Usually isotopes are described by their half-lives ($T_{1/2}$ = time required for one-half of the isotope to decay).

A very useful equation which expresses $T_{1/2}$ is

$$\ln \tfrac{1}{2} = \lambda T_{1/2}$$

or

$$\boxed{T_{1/2} = \frac{0.693}{\lambda}}$$

From this equation the radioactive decay constant (λ) may be calculated and substituted in an earlier equation to obtain

$$\log \frac{N}{N_0} = -\frac{0.693}{T_{1/2}/2.3} \times t$$

C. MEASUREMENT OF ISOTOPES

Radioactivity is commonly measured by the ionization produced when radiation particles interact with the planetary electrons of the atoms of a particular counting gas. The ion pair produced by the radiation consists of the ejected electron and the residual, positively charged atom. The electron can be ejected from any of the orbitals, not necessarily only from the outer valence shell. The ions, of course, have a great tendency to recombine. Quantitative measurement then depends on separating and counting the ions before appreciable recombination occurs. Separation is accomplished by allowing the ionization to occur in an electric field so that the electrons are accelerated very rapidly toward the anode, and so that the heavy, positive ions move slowly toward the cathode.

The electric field can be controlled so that: only the electrons liberated by the passage of the ionizing particle are measured; or the electrons produced by radiation particles are accelerated and cause ejection of other electrons from other atoms of gas, resulting in further multiplication of the original ionization. In the first method, single electrons released as a result of collision of an ionizing particle with a gas atom are measured in an ionization chamber connected to a sensitive electrometer. This method measures the total ionization in the volume of the chamber. In the second method, showers of electrons produced by multiplication of a single ionization event are measured in a Geiger-Müller counter or a proportional counter attached to a scaling circuit and a mechanical register. The general relationship between ionization and potential of the electric field is presented in Figure 65. Ionization chambers are, in general, less sensitive than Geiger-Müller or proportional counters, but the ionization chambers have a lower background correction.

1. Proportional Counters

As the potential of the electric field is increased above the region at which ionization is constant, the electrons produced by the ionizing current are accelerated at such high velocities that they themselves cause ionizations. Consequently, the number of electrons reaching the anode represents a multiplication of the effect of the ionizing particle.

FIGURE 65. The relationship between electric potential and the amount of ionization caused by a radioactive particle.

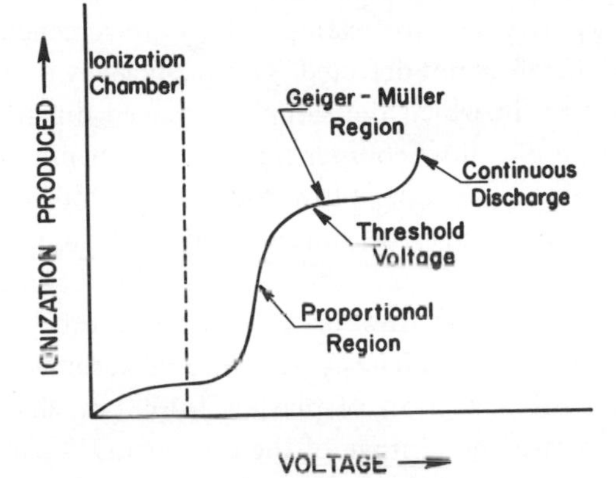

In this region the number of electrons is still proportional to the number of ion pairs produced. The proportionality factor depends on whether a small number of ion pairs are produced (β particles) or several thousand (α particles). Originally, proportional counters were used to count α particles at relatively low voltages, with a consequently low proportionality factor. With considerably increased voltage, proportional counters can be used with great efficiency to count β particles.

Both proportional counters and Geiger-Müller counters require the presence of a specific polyatomic "counting gas" (for example, methane or 1% butane in helium) in order to provide the proper atoms for multiple ionizations. The efficiency of these counters at any given voltage is a function of the nature of the counting gas used.

2. Geiger-Müller Counters

Beyond the proportional region, there is a transition region at which proportionality decreases, and further beyond there is a zone at which the number of electrons reaching the anode is approximately the same whether the original ionizing particle produces only one ion pair or several thousand within the sensitive volume. This is the Geiger-Müller region, or plateau. The counters used to detect showers of ions (Townsend avalanches) generated by particles hitting polyatomic gas molecules in this voltage region are Geiger-Müller tubes. These tubes operate at high voltages.

The usual Geiger-Müller tube consists of a cylindrical cathode at ground potential and a fine, high-voltage central wire, which is the anode. The tube is filled with a counting gas mixture held in the Geiger tube by a window of mica (1–3 mg/cm^2). Therefore radioactive particles must pass through the window in order to be detected. High-energy β particles (for example, ^{32}P) usually penetrate such end window counters, but many low-energy particles (for example, ^{14}C) are screened out by the window and are therefore not detected. This inefficiency is overcome in the gas flow counter, in which the sample is placed directly in the tube. The construction of a flow counter includes an opening through which the counting gas can escape. Flow counters, therefore, require preflushing before counting may be started, and a continual flow of counting gas during operation, hence the name gas flow counter.

For many years the thin window or flow window tube has been commonly used because of its efficiency and ease of automation. This tube has an extremely thin window of plastic (1/4000 in. thick). This thin window allows a large percentage of the low-energy β particles to penetrate and therefore is more efficient than the mica end window tube. Yet the extreme thinness of the window allows a slow leakage of counting gas. Hence the thin window tubes are also flushed with counting gas during operation.

3. Scintillation Counter

Since certain organic phosphors emit photons when hit by charged particles, the number of radioactive disintegrations of an added sample can be counted by measuring the number of light flashes produced. This extremely sensitive method can be used accurately for samples of low specific activity and for low-energy β particles (for example, from tritium) which are difficult to detect efficiently by other methods. Today the scintillation counter is most commonly used for the measurement of radioactive isotopes in biology. Some of the reasons for using the scintillation counter are:

(1) High efficiency for most isotopes.

(2) The low cost of tritium in comparison to ^{14}C compounds makes it a desirable isotope.

(3) The possibility of counting two different isotopes within the same sample (double labeling).

GENERAL REFERENCES

Clark, John M. *Experimental Biochemistry*. W. H. Freeman and Co., San Francisco and London, 1964.

Kamen, M. D. *Isotopic Tracers in Biology*, 3rd rev. ed., Academic Press, New York, 1957.

Steinberg, D., and S. Udenfriend. The Measurement of Radioisotopes. In *Methods in Enzymology*. Vol. IV, Eds. S. P. Colowick and N. O. Kaplan. Academic Press, New York. 1957, p. 425.

APPENDIX 10 BUFFERS COMMONLY USED IN BIOLOGY

A. HYDROCHLORIC ACID–POTASSIUM CHLORIDE BUFFER (1)

Stock solutions.

A: 0.2 *M* solution of KCl (14.91 g in 1000 ml)

B: 0.2 *M* HCl

50 ml of A, *x* ml of B, diluted to a total of 200 ml

x	pH	*x*	pH
97.0	1.0	20.6	1.7
78.0	1.1	16.6	1.8
64.5	1.2	13.2	1.9
51.0	1.3	10.6	2.0
41.5	1.4	8.4	2.1
33.3	1.5	6.7	2.2
26.3	1.6		

B. GLYCINE-HCL BUFFER (2)

Stock solutions.

A: 0.2 *M* solution of glycine (15.01 g in 1000 ml)

B: 0.2 *M* HCl

50 ml of A, *x* ml of B, diluted to a total of 200 ml

x	pH	*x*	pH
5.0	3.6	16.8	2.8
6.4	3.4	24.2	2.6
8.2	3.2	32.4	2.4
11.4	3.0	44.0	2.2

C. PHTHALATE–HYDROCHLORIC ACID BUFFER (1)

Stock solutions.

A: 0.2 *M* solution of potassium acid phthalate (40.84 g in 1000 ml)

B: 0.2 *M* HCl

50 ml of A, *x* ml of B, diluted to a total of 200 ml

x	pH	*x*	pH
46.7	2.2	14.7	3.2
39.6	2.4	9.9	3.4
33.0	2.6	6.0	3.6
26.4	2.8	2.63	3.8
20.3	3.0		

D. ACONITATE BUFFER (3)

Stock solutions.

A: 0.5 *M* solution of aconitic acid (87.05 g in 1000 ml)

B: 0.2 *M* NaOH

20 ml of A, *x* ml of B, diluted to a total of 200 ml

x	pH	*x*	pH
15.0	2.5	83.0	4.3
21.0	2.7	90.0	4.5
28.0	2.9	97.0	4.7
36.0	3.1	103.0	4.9
44.0	3.3	108.0	5.1
52.0	3.5	113.0	5.3
60.0	3.7	119.0	5.5
68.0	3.9	126.0	5.7
76.0	4.1		

E. CITRATE BUFFER (4)

Stock solutions.

A: 0.1 *M* solution of citric acid (21.01 g in 1000 ml)

B: 0.1 *M* solution of sodium citrate (29.41 g $C_6H_5O_7Na_3 \cdot 2\,H_2O$ in 1000 ml)

x ml of A, *y* ml of B, diluted to a total of 100 ml

x	*y*	pH	*x*	*y*	pH
46.5	3.5	3.0	23.0	27.0	4.8
43.7	6.3	3.2	20.5	29.5	5.0
40.0	10.0	3.4	18.0	32.0	5.2
37.0	13.0	3.6	16.0	34.0	5.4
35.0	15.0	3.8	13.7	36.3	5.6
33.0	17.0	4.0	11.8	38.2	5.8
31.5	18.5	4.2	9.5	41.5	6.0
28.0	22.0	4.4	7.2	42.8	6.2
25.5	24.5	4.6			

F. ACETATE BUFFER (5)

Stock solutions.

A: 0.2 *M* solution of acetic acid (11.55 ml in 1000 ml)

B: 0.2 *M* solution of sodium acetate (16.4 g of $C_2H_3O_2Na$ or 27.2 g of $C_2H_3O_2Na \cdot 3\ H_2O$ in 1000 ml)

x ml of A, *y* ml of B, diluted to a total of 100 ml

x	*y*	pH	*x*	*y*	pH
46.3	3.7	3.6	20.0	30.0	4.8
44.0	6.0	3.8	14.8	35.2	5.0
41.0	9.0	4.0	10.5	39.5	5.2
36.8	13.2	4.2	8.8	41.2	5.4
30.5	19.5	4.4	4.8	54.2	5.6
25.5	24.5	4.6			

G. CITRATE-PHOSPHATE BUFFER (6)

Stock solutions.

A: 0.1 *M* solution of citric acid (19.21 g in 1000 ml)

B: 0.2 *M* solution of dibasic sodium phosphate (53.65 g of $Na_2HPO_4 \cdot 7\ H_2O$ or 71.7 g of $Na_2HPO_4 \cdot 12\ H_2O$ in 1000 ml)

x ml of A, *y* ml of B, diluted to a total of 100 ml

x	*y*	pH	*x*	*y*	pH
44.6	5.4	2.6	24.3	25.7	5.0
42.2	7.8	2.8	23.3	26.7	5.2
39.8	10.2	3.0	22.2	27.8	5.4
37.7	12.3	3.2	21.0	29.0	5.6
35.9	14.1	3.4	19.7	30.3	5.8
33.9	16.1	3.6	17.9	32.1	6.0
32.3	17.7	3.8	16.9	33.1	6.2
30.7	19.3	4.0	15.4	34.6	6.4
29.4	20.6	4.2	13.6	36.4	6.6
27.8	22.2	4.4	9.1	40.9	6.8
26.7	23.3	4.6	6.5	43.6	7.0
25.2	24.8	4.8			

H. SUCCINATE BUFFER (3)

Stock solutions.

A: 0.2 *M* solution of succinic acid (23.6 g in 1000 ml)

B: 0.2 *M* NaOH

25 ml of A, *x* ml of B, diluted to a total of 100 ml

x	pH	*x*	pH
7.5	3.8	26.7	5.0
10.0	4.0	30.3	5.2
13.3	4.2	34.2	5.4
16.7	4.4	37.5	5.6
20.0	4.6	40.7	5.8
23.5	4.8	43.5	6.0

I. PHOSPHATE–SODIUM HYDROXIDE BUFFER (1)

Stock solutions.

A: 0.2 *M* solution of potassium acid phosphate (40.84 g in 100 ml)

B: 0.2 *M* NaOH

50 ml of A, *x* ml of B, diluted to a total of 200 ml

x	pH	*x*	pH
3.7	4.2	30.0	5.2
7.5	4.4	35.5	5.4
12.2	4.6	39.8	5.6
17.7	4.8	43.0	5.8
23.9	5.0	45.5	6.0

J. MALEATE BUFFER (7)

Stock solutions.

A: 0.2 *M* solution of acid sodium maleate (8 g of NaOH, 23.2 g of maleic acid, or 19.6 g of maleic anhydride in 1000 ml)

B: 0.2 *M* NaOH

50 ml of A, *x* ml of B, diluted to a total of 200 ml

x	pH	*x*	pH
7.2	5.2	33.0	6.2
10.5	5.4	38.0	6.4
15.3	5.6	41.6	6.6
20.8	5.8	44.4	6.8
26.9	6.0		

K. CACODYLATE BUFFER (8)

Stock solutions.

A: 0.2 *M* solution of sodium cacodylate (42.8 g of $Na(CH_3)_2AsO_2 \cdot 3\,H_2O$ in 1000 ml)

B: 0.2 *M* HCl

50 ml of A *x* ml of B, diluted to a total of 200 ml

x	pH	*x*	pH
2.7	7.4	29.6	6.0
4.2	7.2	34.8	5.8
6.3	7.0	39.2	5.6
9.3	6.8	43.0	5.4
13.3	6.6	45.0	5.2
18.3	6.4	47.0	5.0
23.8	6.2		

L. PHOSPHATE BUFFER (2)

Stock solutions.

A: 0.2 *M* solutions of monobasic sodium phosphate (27.8 g in 1000 ml)

B: 0.2 *M* solution of dibasic sodium phosphate (53.65 g of $Na_2HPO_4 \cdot 7\ H_2O$ or 71.7 g of $Na_2HPO_4 \cdot 12\ H_2O$ in 1000 ml)

x ml of A, *y* ml of B, diluted to a total of 200 ml

x	*y*	pH	*x*	*y*	pH
93.5	6.5	5.7	45.0	55.0	6.9
92.0	8.0	5.8	39.0	61.0	7.0
90.0	10.0	5.9	33.0	67.0	7.1
87.7	12.3	6.0	28.0	72.0	7.2
85.0	15.0	6.1	23.0	77.0	7.3
81.5	18.5	6.2	19.0	81.0	7.4
77.5	22.5	6.3	16.0	84.0	7.5
73.5	26.5	6.4	13.0	87.0	7.6
68.5	31.5	6.5	10.5	90.5	7.7
62.5	37.5	6.8	8.5	91.5	7.8
56.5	43.5	6.7	7.0	93.0	7.9
51.0	49.0	6.8	5.3	94.7	8.0

M. TRIS(HYDROXYMETHYL)AMINOMETHANE-MALEATE (TRIS-MALEATE) BUFFER (9, 10)

Stock solutions.

A: 0.2 *M* solution of Tris acid maleate [2.4 g of Tris(hydroxymethyl)aminomethane plus 23.2 g of maleic acid or 19.6 g of maleic anhydride in 1000 ml]

B: 0.2 M NaOH

50 ml of A, x ml of B, diluted to a total of 200 ml

x	pH	x	pH
7.0	5.2	48.0	7.0
10.8	5.4	51.0	7.2
15.5	5.6	54.0	7.4
20.5	5.8	58.0	7.6
26.0	6.0	63.5	7.8
31.5	6.2	69.0	8.0
37.0	6.4	75.0	8.2
42.5	6.6	81.0	8.4
45.0	6.8	86.5	8.6

N. BARBITAL BUFFER (11)

Stock solutions.

A: 0.2 M solution of sodium barbital (Veronal) (41.2 g in 1000 ml)

B: 0.2 M HCl

50 ml of A, x ml of B, diluted to a total of 200 ml

x	pH	x	pH
1.5	9.2	22.5	7.8
2.5	9.0	27.5	7.6
4.0	8.8	32.5	7.4
6.0	8.6	39.0	7.2
9.0	8.4	43.0	7.0
12.7	8.2	45.0	6.8
17.5	8.0		

Solutions more concentrated than 0.05 M may crystallize on standing, especially in the cold.

O. TRIS(HYDROXYMETHYL)AMINOMETHANE (TRIS-HCL) BUFFER (9)

Stock solutions.

A: 0.2 M solution of Tris(hydroxymethyl)aminomethane (24.2 g in 1000 ml)

B: 0.2 M HCl

50 ml of A, x ml of B, diluted to a total of 200 ml

x	pH	x	pH
5.0	9.0	26.8	8.0
8.1	8.8	32.5	7.8
12.2	8.6	38.4	7.6
16.5	8.4	41.4	7.4
21.9	8.2	44.2	7.2

P. BORIC ACID–BORAX BUFFER (12)

Stock solutions.

A: 0.2 *M* solution of boric acid (12.4 g in 1000 ml)

B: 0.05 *M* solution of borax (19.05 g in 1000 ml; 0.2 *M* in terms of sodium borate)

50 ml of A, *x* ml of B, diluted to a total of 200 ml

x	pH	*x*	pH
2.0	7.6	22.5	8.7
3.1	7.8	30.0	8.8
4.9	8.0	42.5	8.9
7.3	8.2	59.0	9.0
11.5	8.4	83.0	9.1
17.5	8.6	115.0	9.2

Q. 2-AMINO-2-METHYL-1,3-PROPANEDIOL (AMMEDIOL) BUFFER (13)

Stock solutions.

A: 0.2 *M* solution of 2-amino-2-methyl-1,3-propanediol (21.03 g in 1000 ml)

B: 0.2 *M* HCl

50 ml of A, *x* ml of B, diluted to a total of 200 ml

x	pH	*x*	pH
2.0	10.0	22.0	8.8
3.7	9.8	29.5	8.6
5.7	9.6	34.0	8.4
8.5	9.4	37.7	8.2
12.5	9.2	41.0	8.0
16.7	9.0	43.5	7.8

R. GLYCINE-NaOH BUFFER (2)

Stock solutions.

A: 0.2 *M* solution of glycine (15.01 g in 1000 ml)

B: 0.2 *M* NaOH

50 ml of A, *x* ml of B, diluted to a total of 200 ml

x	pH	*x*	pH
4.0	8.6	22.4	9.6
6.0	8.8	27.2	9.8
8.8	9.0	32.0	10.0
12.0	9.2	38.6	10.4
16.8	9.4	45.5	10.6

S. BORAX-NaOH BUFFER (1)

Stock solutions.

A: 0.05 *M* solution of borax (19.05 g in 1000 ml; 0.02 *M* in terms of sodium borate)

B: 0.2 *M* NaOH

50 ml of A, *x* ml of B, diluted to a total of 200 ml

x	pH	*x*	pH
0.0	9.28	29.0	9.7
7.0	9.35	34.0	9.8
11.0	9.4	38.0	9.9
17.6	9.5	43.0	10.0
23.0	9.6	46.0	10.1

T. CARBONATE-BICARBONATE BUFFER (14)

Stock solutions.

A: 0.2 *M* solution of anhydrous sodium carbonate (21.2 g in 1000 ml)

B: 0.2 *M* solution of sodium bicarbonate (16.8 g in 1000 ml)

x ml of A, *y* ml of B, diluted to a total of 200 ml

x	*y*	pH	*x*	*y*	pH
4.0	46.0	9.2	27.5	22.5	10.0
7.5	42.5	9.3	30.0	20.0	10.1
9.5	40.5	9.4	33.0	17.0	10.2
13.0	37.0	9.5	35.5	14.5	10.3
16.0	34.0	9.6	38.5	11.5	10.4
19.5	30.5	9.7	40.5	9.5	10.5
22.0	28.0	9.8	42.5	7.5	10.6
25.0	25.0	9.9	45.0	5.0	10.7

REFERENCES

1. W. M. Clark and H. A. Lubs. *J. Bacteriol.* 2: 1 (1917).
2. S. P. L. Sorensen. *Biochem. Z. 21*, 131 (1909); 22: 352 (1909).
3. G. Gamori. Unpublished data.
4. R. D. Lillie. *Histopathologic Technique*. Blakiston, Philadelphia and Toronto. 1948.
5. G. S. Walpole. *J. Chem. Soc.* 105: 2501 (1914).
6. T. C. McIlvaine. *J. Chem.* 49: 183 (1921).
7. J. W. Temple. *J. Am. Chem. Soc.* 51: 1754 (1929).
8. M. Plumel. *Bull. Soc. Chim. Biol.* 30: 129 (1949).
9. Sigma Chemical Co., St. Louis, Mo.; Matheson Coleman and Bell,

East Rutherford, N.J. (Buffer grade Tris can be obtained from these companies).
10. G. Gomori. *Proc. Soc. Exp. Biol. Med.* 68: 354 (1948).
11. L. Michaelis. *J. Biol. Chem.* 87: 33 (1930).
12. W. Holmes. *Anat. Rec.* 86: 163 (1943).
13. G. Gomori. *Proc. Soc. Exp. Biol. Med.* 62: 33 (1946).

INDEX